The Economic Dimensions of Crime

Also by Nigel G. Fielding

ACTIONS AND STRUCTURE

COMMUNITY POLICING

COMPUTER ANALYSIS AND QUALITATIVE RESEARCH

INVESTIGATING CHILD SEXUAL ABUSE

JOINING FORCES

LINKING DATA

THE NATIONAL FRONT

NEGOTIATING NOTHING

THE POLICE AND SOCIAL CONFLICT

PROBATION PRACTICE

USING COMPUTERS IN QUALITATIVE RESEARCH

Also by Alan Clarke

EVALUATION RESEARCH: An Introduction to Principles, Methods and Practice

The Economic Dimensions of Crime

Edited by

Nigel G. Fielding
Professor of Sociology and
Co-Director, Institute of Social Research
University of Surrey

Alan Clarke
Lecturer in Sociology
University of Surrey

Robert Witt
Senior Lecturer in Economics
University of Surrey

First published in Great Britain 2000 by
MACMILLAN PRESS LTD
Houndmills, Basingstoke, Hampshire RG21 6XS and London
Companies and representatives throughout the world

A catalogue record for this book is available from the British Library.

ISBN 0–333–76038–7

First published in the United States of America 2000 by
ST. MARTIN'S PRESS, LLC,
Scholarly and Reference Division,
175 Fifth Avenue, New York, N.Y. 10010

ISBN 0–312–23161–X

Library of Congress Cataloging-in-Publication Data
The economic dimensions of crime / edited by Nigel G. Fielding, Alan Clarke,
Robert Witt.
p.   cm.
Includes bibliographical references and index.
ISBN 0–312–23161–X
1. Crime—Economic aspects. 2. Criminal justice, Administration of—Economic
aspects. I. Fielding, Nigel. II. Clarke, Alan, 1951- III. Witt, Robert.
HV6171 .E28   2000
364—dc21
99–088101

This book is printed on paper suitable for recycling and made from fully managed and sustained
forest sources.

10   9   8   7   6   5   4   3   2   1
09   08   07   06   05   04   03   02   01   00

Printed and bound in Great Britain by
Antony Rowe Ltd, Chippenham, Wiltshire

Contents

List of Tables   vii

List of Figures   viii

Notes on the Contributors   x

An Introduction to the Economic Dimensions of
Crime and Punishment   1
    Nigel G. Fielding, Alan Clarke and Robert Witt

**PART I   THE ECONOMIC ANALYSIS OF CRIME AND
PUNISHMENT**

1   Crime and Punishment: an Economic Approach   15
      Gary S. Becker

2   The Economics of Crime   70
      Antony W. Dnes

3   Economists, Crime and Punishment   83
      David Pyle

4   Conspiracy among the Many: the Mafia in Legitimate
   Industries   100
      Diego Gambetta and Peter Reuter

5   Towards an Economic Approach to Crime and
   Prevention   122
      **Graham Farrell, Sharon Chamard, Ken Clark
      and Ken Pease**

**PART II   CRIME AND THE LABOUR MARKET:
ECONOMIC AND STRUCTURAL FACTORS**

6   Crime and Consumption   142
      Simon Field

7   Crime and the Labour Market   150
      Richard B. Freeman

8   Work and Crime: an Exploration Using Panel Data   177
      Ann D. Witte and Helen Tauchen

9   'The Devil Finds Work for Idle Hands to Do':
    the Relationship between Unemployment and
    Delinquency   193
    Gerald Prein and Lydia Seus

10 Crime, Unemployment and Deprivation   210
    Alan Clarke, Nigel G. Fielding and Robert Witt

**PART III   MODELLING THE SYSTEM-WIDE COSTS OF
CRIMINAL JUSTICE POLICIES AND PROGRAMMES**

11 Modelling the Cost of Crime   226
    **Allen K. Lynch, Todd Clear and
    David W. Rasmussen**

12 Auditing Criminal Justice   239
    Joanna Shapland

Annotated Further Readings   251

Index   253

List of Tables

1.1	Economic costs of crimes	17
1.2	Probability of conviction and average prison term for several major felonies, 1960	31
2.1	Cost of reducing property crime by 1 per cent	75
3.1	Stealing as a dominant strategy	85
4.1	Conditions favouring the emergence of mafia-controlled cartels	112
5.1	Sixteen techniques of situational crime prevention	129
8.1	Results for the probability of offending	184
9.1	Ordinal models for violent delinquency	197
10.1	Regional crime and unemployment rates, United Kingdom, 1979–93	216
11.1	Definitions of variables used in housing market analysis	231
11.2	Hedonic estimates of the willingness to pay for public safety	233

List of Figures

1.1 Marginal cost and marginal revenue by numbers of offences   27
1.2 Effect of marginal damages on marginal cost of changing offences   30
1.3 Marginal costs of apprehension and convictions   32
1.4 Effect of reduced elasticity of offences   33
3.1 The opportunity locus for legitimate and criminal activity   89
3.2 Optimum allocation of time between legitimate and criminal activity   90
3.3 Changes in the certainty of punishment and the indifference curve   91
3.4 Recorded offences of burglary, England and Wales, 1946–96   94
3.5 Long-run/short-run crime–economy relationships   95
3.6 Annual percentage changes in real GDP and recorded burglary, 1986/87 to 1995/96   95
5.1 (a) Allocating time to crime: the marginal utility curves of time spent offending and time spent not offending; (b) a preference for crime? Different marginal utility curves results in offenders and non-offenders   125
5.2 Risk and crime model   127
5.3 Increasing (time and effort and) risks of detection reduces criminal opportunities   130
5.4 Reducing rewards or removing excuses reduces the demand for offences by offenders   132
5.5 Repeat victimisation committed with reduced risk uncertainty   134
5.6 Offending under uncertainty of both risk and rewards   135
6.1 Property crime and consumption   143
7.1 Prison and jail populations in the United States, 1947–92   152
7.2 Uniform Crime Reporting index per 100 000, 1947–92   152
7.3 Victimisations per 100 000, 1973–92   153
7.4 Crimes and confined population per adult male, 1977–92   155
7.5 Victimisation and confined population per adult male, 1977–92   156
7.6 Crimes per adult male and unemployment, 1948–92   158

7.7   Property crime versus income inequality in
       127 metropolitan areas in the United States, 1979   162

Notes on the Contributors

Gary S. Becker is Professor of Economics, University of Chicago, Chicago, Illinois, USA.

Sharon Chamard (M.Sc., Rutgers, 1997) is a doctoral student at the School of Criminal Justice at Rutgers University, New Brunswick, New Jersey, USA.

Ken Clark (M.Sc., Manchester, 1990) is Lecturer in Econometrics at the School of Economic Studies at Manchester University, England, and a sometime Visiting Fellow at the Center for Crime Prevention Studies in the School of Criminal Justice at Rutgers University, New Brunswick, New Jersey, USA.

Alan Clarke is Lecturer in Sociology at the University of Surrey, Guildford, England. He was educated at the universities of Middlesex, Hull and Nottingham.

Todd Clear received his Ph.D. from the State University of New York and serves as Professor and Associate Dean in the School of Criminology and Criminal Justice, Florida State University, Florida, USA.

Antony W. Dnes is Professor of Economics and Associate Dean for Research, University of Hertfordshire, England.

Graham Farrell (Ph.D., Manchester, 1994) is Deputy Research Director of the Police Foundation, Washington, DC, USA, and Research Associate of the Applied Criminology Group of the University of Huddersfield, England.

Simon Field is former Head of Economics of the Home Office Research Development and Statistics Directorate, London, England.

Nigel G. Fielding is Professor of Sociology and co-Director of the Institute of Social Research at the University of Surrey, Guildford, England. He was educated at the universities of Sussex and Kent, and at the London School of Economics.

Richard B. Freeman is Program Director (Labor Studies), National Bureau of Economic Research, Cambridge, Massachusetts, USA.

Diego Gambetta is with the Department of Applied Social Studies, University of Oxford, England.

Allen K. Lynch received his Ph.D. from Florida State University and is Assistant Professor of Economics at the University of North Florida, Florida, USA.

Ken Pease (Ph.D., Manchester, 1971) is acting Head of the Policing and Researching Crime Unit of the Home Office, London, England.

Gerald Prein is a senior research scientist at the Special Research Centre and a lecturer in sociological methodology and statistics at the University of Bremen, Germany.

David Pyle is Professor of Applied Economics and Dean of the Faculty of Social Sciences at the University of Leicester, England. He was educated at the universities of York and Bristol.

David W. Rasmussen, Ph.D. from Washington University, St Louis, is Professor of Economics and Director of the DeVoe L. Moore Center for the Study of Critical Issues in Economic Policy and Government at Florida State University, Florida, USA.

Peter Reuter is with the School of Public Affairs, University of Maryland, College Park, Maryland, USA.

Lydia Seus is a senior research scientist at the Special Research Centre and a lecturer in sociology and criminology at the University of Bremen, Germany.

Joanna Shapland is Professor of Criminal Justice and Director of the Institute for the Study of the Legal Profession in the Department of Law at the University of Sheffield, England.

Helen Tauchen is Professor of Economics, University of North Carolina, Chapel Hill, North Carolina, USA.

Robert Witt is Senior Lecturer in Economics at the University of Surrey, Guildford, England. He was educated at Kingston University, the University of Wales (Bangor) and the University of Essex.

Ann D. Witte is Professor of Economics, Florida International University, Florida, USA.

An Introduction to the Economic Dimensions of Crime and Punishment

Nigel G. Fielding, Alan Clarke and Robert Witt

We hope this book will raise the profile in criminology of economic perspectives on crime and criminal justice. Its contents include a mixture of previously published exemplars and original contributions reporting contemporary work in key areas of the economic analysis of crime, which we have sought to weld into a whole by our editorial contributions. The chapters are by established figures in several fields, including economics, psychology, sociology and law.

The economic dimension of crime is an underdeveloped perspective in criminology (Fiorentini and Peltzman, 1995). Even in the USA the widely regarded keynote contribution was itself made just over 30 years ago (Becker, 1968; reprinted in this volume as Chapter 1). This relative underdevelopment of economic input to criminology contrasts with the research questions, analytic and applied, which criminology has lately come to address. Further, there is a considerable body of applied and theoretical research by economists and those taking an economic perspective on crime and criminal justice (e.g. Wilson, 1983) which could profitably be brought to wider notice in criminology. We hope that our book promotes creative exchange between the economics discipline, other social sciences touching on economic perspectives, and their application to problems of crime and criminal justice.

Bearing in mind that the field is somewhat further advanced in the USA, the book includes previously published exemplars by American economic criminologists (reproduced as they originally appeared, save for minor editing to ensure stylistic consistency and to complete the bibliographic references), and, to bring these perspectives up to date and show how they have been applied, original contributions from the USA, Britain and Germany. The scope of the book is not confined to economic explanations of trends in crime or individual agents' allocative choice between legal and criminal activities but extends to issues surrounding the costs of crime control, crime prevention and so on. The main sections of the book

1

are Part I: 'The Economic Analysis of Crime and Punishment'; Part II: 'Crime and the Labour Market: Economic and Structural Factors' and Part III: 'Modelling the System-wide Costs of Criminal Justice Policies and Programmes'. Criminology's relative neglect of economic perspectives is largely because criminologists and criminal justice researchers are unfamiliar with economic concepts and methods. Accordingly, this initial essay points up the principal elements of economic approaches to crime and criminals, and signposts each chapter's place in the overall economics context. Between chapters there is a bridging discussion, highlighting the connections between the chapter just ended and the chapter to come. There is also an annotated list of further readings.

Economists approach crime and punishment as a special case of the more general economic theory of choice. A key feature is the notion of utility; judgements are made of the gain to be realised (the 'expected utility') from a particular course of action. If the expected utility of a given activity exceeds that of another activity it is predicted that an individual will prefer the given activity. Notions of 'supply' and 'demand' are also relevant; a key consideration here is whether demand is elastic or inelastic. Matters of crime and punishment are understood as resulting from choices made between criminal and non-criminal courses of action informed by weighing their rewards and risks. This heuristic accounts for individual decision-making and the socioeconomic context in which it occurs. The first section of the book seeks to illustrate what can be done by working from these assumptions.

All the social sciences construct models of human behaviour in their own characteristic ways but economics has a particular preference for the formal, or mathematical, model. Several contributions in the first section exemplify the use of such models. These proceed from the first contribution, Gary Becker's seminal essay. Although the essay touches on many aspects of crime and punishment, Becker's central interest is in establishing what degree of legal regulation (law enforcement plus punishment) should be committed to enforcing different laws. The social loss caused by different infractions is put in relation to the likely effects and costs of different punishments (or criminal justice system responses). The approach sidesteps the aetiological concerns that mark the approach of other disciplines in favour of testing models based on the cost of apprehending and convicting offenders, types of punishment, and the responses offenders make to changes in law enforcement. A particular concern is to evaluate responses to crime based on increasing the probability of arrest and conviction versus responses based on increasing the severity of punishments. The matter brings into play the importance of offenders' attitudes towards risk.

In Chapter 2 Antony Dnes builds on Becker's approach, exploring the deterrence hypothesis which suggests criminal activity can be reduced by increasing the costs and decreasing the benefits of crime. Dnes emphasizes the need to establish which factors in the model most significantly influence criminal behaviour. Despite problems of measurement, we are in a position to say that results vary for different kinds of offence: they display different elasticity relative to the clear-up rate (for example, non-sexual violent offences compared to property-related offences). Dnes also maintains that, while probability of conviction may show stronger deterrent effects than severe punishment, when cost is taken into account the latter is a more efficient option.

Another thread of Becker's approach informs David Pyle's interest in the impact of economic activity on participation in crime (see Chapter 3). Pyle construes criminal participation as a labour supply decision, so that the decision to participate is seen as related to potential earnings from lawful employment, the returns from crime, and the probability of unemployment. Pyle shows how the trade-offs between these factors can be modelled, affording a better basis for predicting under what conditions individuals are likely to commit crimes. But the focus is not purely on individual allocative choice. Pyle introduces us to some of the recent evidence for links between crime and the state of the economy, and brings in the importance of the temporal dimension when we are assessing the effects of different structural conditions.

Pyle's interest in the relationship between crime and macroeconomic trends raises a major criminological debate in which economics has played an important part, the role of unemployment as a factor motivating criminal involvement. We will return to this matter, but there is another criminological preoccupation to which economic perspectives have contributed, the (contested) growth and power of organized crime. Diego Gambetta (1994) and Peter Reuter (1983) have separately carried out case study research on syndicated criminal organisations. Gambetta and Reuter see organised crime as a means of criminal dispute settlement allowing cartel-type economic arrangements to continue functioning through contract enforcement. They draw on the economic theory of the firm to demonstrate a number of counter-intuitive findings which do much to explain the relations between organised crime and the business world. Among them is the idea that, rather than straightforward victims, restrained by the violence of syndicate enforcers, businesses gain much from being 'coerced'. Gambetta and Reuter suggest that the racketeers effectively mobilise and maintain trade allocation agreements between companies, and that this can keep in business companies which might otherwise succumb to the rigours of competition. Other scholars have also pursued an economic approach to organised crime; for example, Van Duyne (1993, 1996) has examined organised crime in northwest Europe as an economy based on cross-border crime trade, and Levi has addressed European organised crime as a form of trade adapted to 'a large diversity of economies, extensive economic regulations, many loosely controlled borders to cross, and relatively small jurisdictions' (Levi, 1998: 338–9). Such research leads us away from Godfather stereotypes towards an altogether more troubling alliance of mafiosi, politicians and business. The Gambetta and Reuter chapter (Chapter 4) also demonstrates the way that economics is used in tandem with knowledge from other disciplines.

Having brought in the economic contribution to controversial topics in criminology, the first section closes with a chapter addressing one of the liveliest growth points in contemporary criminal justice, crime prevention. Farrell, Chamard, Clark and Pease (Chapter 5) give an idea not only of how economics can contribute to crime prevention but how criminological and crime prevention *theory* can be integrated by drawing on ideas from economics. Working from the decision heuristic presented in the first three chapters, they show why this heuristic makes for repeat victimization of known targets, an understanding which can be used in designing crime prevention interventions. Thus, the economic perspective based on individual choice between alternative courses of action allows us insight into a number of facets of crime and punishment. In the next section our attention turns to the relationship between crime and structural factors; in particular, the labour market.

Economics and common sense agree that it is to be expected that the state of the economy – local, regional and national – will have an effect on crime rates. While plausible, reflection impresses upon us how complicated the intervening factors are between these levels of the economy and the process of individual decision-making through which economics understands crime. For example, whether welfarist or market-oriented, all modern states attempt to buffer the harshest effects of economic crises on their citizens. We might also mention, as a complicating factor for any model, the unknown but substantial 'informal economy'. Both the welfare provision and informal economic aspects of national economies intrude on models allowing only a rigidly bipolar trade-off between 'legitimate employment' and 'crime'. As we shall see, even investigating the most straightforward propositions requires long-term trend data. Tracing the actual effect of unemployment on criminal participation requires considerable methodological sophistication. If we want to trace the effects of economic conditions at a regional level – which can be a useful means of evaluating relationships by comparing effects within relatively 'closed' systems – we still encounter measurement problems. We might expect that the merit of such an approach is consistency of legal definitions and data sets. Regrettably this assumption is often confounded, for example, by differences between police forces in crime-recording practices.

Simon Field's chapter (Chapter 6) exemplifies a trend analysis taking the long view. He looks at trends in the economy and in crime rates in England and Wales over a time frame of most of the twentieth century. Field has something to say about unemployment but it is not what common sense might expect. He maintains that, when we take account of the temporal factor, we do not see as direct an impact of unemployment on crime as we might expect. A stronger relationship can be established between per-capita real personal consumption (an index of relative wealth) and crime. The more prosperity, the slower the growth in the rate of property crime; it may even fall. But 'personal crime' does not relate to the economy in the same way. Moreover, while similar trends can be seen in several other countries, these trends cannot be seen in all. Further, when examining these macro-level trends over long time periods, problems of measurement become considerable, and if we seek to validate our analysis by international comparison we move into real technical complexity. Even if we take one country's crime data, for instance, the way that crime is officially recorded changes from time to time, and the further back we go the less likely we will pick up all the changes.

It will be recalled that Becker's deterrence hypothesis, and its interpretation by Dnes, suggests crime rates should be responsive to severity of sentencing. Yet, as Richard Freeman observes (Chapter 7), high crime rates were suffered in the USA throughout the 1980s although it was a time of significantly harsher sentencing. Freeman considers whether other incentives to commit crime could be implicated, like the deteriorating earnings of unskilled workers, or the widening income gap during the period of 'Reaganomics'. With increasing criminal participation during the period, particularly among males, Freeman suggests economics needs to respond by looking at the 'reverse' of the usual labour market/crime relationships. That is, what effect does crime have on labour market outcomes? What is the effect of criminal participation and a criminal record on future employment? Freeman's position offers succour to those who see low wages, poor vocational training and a weak welfare 'safety net' as criminogenic.

Freeman draws on a variety of sources to pursue his analysis, like the other chapters we have included so far. Witte and Tauchen take a different tack in Chapter 8, reporting a cohort study of a group of young American males. The study pursues the relationship between criminal participation and educational and occupational achievement Freeman raised. While educational achievement and employment showed a nearly identical degree of effect in discouraging criminal participation, it was not a matter of earnings; these lawful modes of adaptation appear to discourage criminal participation in and of themselves. But the educational achievement had to be significant; the effect was not apparent amongst those who had achieved no more than a high school diploma. Witte and Tauchen press the case for more dynamic models which can capture change over time, such as career progression and the process of maturation. Their remarks on dynamic models need to be considered in relation to crime-as-work models like that discussed by Pyle.

Prein and Seus (Chapter 9) make substantial moves to develop such dynamic models, drawing on a longitudinal study of graduates of German vocational education. Such education leads to late entry to the labour market and a relatively high proportion of vocationally qualified employees. With demand for unskilled labour in decline, vocational qualifications are increasingly important for integration into the labour market and increasingly relevant to the distribution of opportunities. Choice between legitimate and illegitimate opportunities is, of course, central to the economic approach, and system-wide constraint in legitimate opportunities would be predictive of more crime. Prein and Seus observe that access to training opportunities is mediated by educational performance, gender and nationality, and those with poor education, females, and children of immigrants, start with relatively high risk factors. These are some contemporary bases of social exclusion to which one might be alert in examining patterns of labour market participation and criminal involvement.

Prein and Seus derive four models with which to test possible relationships between labour market participation and crime. By German standards, the sample featured high levels of unemployment and failure to find work in the field of the respondents' vocational qualification, so any relationship between unemployment and delinquency should be manifest in this study. In fact, Prein and Seus do not find evidence of such a relationship. What they do find are gender-specific patterns whereby relatively more crime is committed by those males who have succeeded in gaining vocationally qualified employment than by the unemployed, such offending being excused or mitigated by employment status when detected, and low levels of criminal involvement amongst unemployed young women accounted for by their recourse to traditional reproductive and domestic roles. Prein and Seus argue that instead of looking for the incidence of delinquency in social groups, we should be looking at which social conditions and life-situations increase or decrease the risk of committing offences. Hence the relevance to the concept of social exclusion.

Aggregate-level studies establish that economic factors impact on crime but we have to go beyond the analysis of rates if we want to understand criminal propensity. There is a distinction between the linking of economic conditions and crime at the macro-level, and the interaction between disadvantage and criminal offending at the individual level. The chapters by Witte and Tauchen, Prein and Seus, and our own chapter (Chapter 10), all address the latter part of this distinction. They all involve some form of disaggregation. In the case of our chapter we used regionally disaggregated data to see if regional differences emerged in the effect of economic factors on crime. An annual time-series and regional cross-sectional data set of offences, unemployment, income variables, socio-economic and sociodemographic data was assembled for ten standard regions of England and Wales. Theft and the handling of stolen goods were examined, both for their largely economic motivation and their frequency of occurrence. We found a positive relationship between the long-term unemployment rate and certain of these offences in data aggregated to regional level; it appeared that the effect of unemployment varied between regions. Across the regions overall, we found that growth in unemployment was matched by increased levels of theft. There is much scope for combining aggregate and individual-level data to more fully explain how economic factors and perceived deprivation feature in crime causation.

In the book's third section we examine the contribution economics can make to understanding criminal justice as a system. In a system the effects of events in one part are not confined to that part but will have effects on other parts of the system too. As a source of predictions that can be modelled and tested against system-level indicators, economics can alert us to unexpected connections. It focuses on the systematic effects of resource allocation, understood using the analogy of the market. One item of information we need when modelling crime and criminal justice is accurate estimates of the victim cost of crime. As well as direct loss occasioned by, say, burglary, there are indirect costs such as increased insurance premiums, the cost of policing and the courts, the lost value of productive work forgone by the householder who has instead to spend time clearing up, and so on. There are also hard-to-quantify costs of upset and anxiety, changed lifestyle (for example, defensive behaviour) and even moving elsewhere. In Chapter 11 Lynch, Clear and Rasmussen apply the market test to the problem by examining the effect of crime on property values. A considerable gap is found between willingness to pay a premium for reduced risk of violent crime and the actual cost of crime at household level; it seems that reduced risk attracts a low premium. However, this is not the case in high crime neighbourhoods, and the authors postulate a contagion effect to account for this. One might make a connection here to Wilson and Kelling's 'broken windows' hypothesis (1982).

The criminal justice system is diverse, multifaceted and surprising. There are a host of agencies involved and at the margins these agencies engage in much work having little to do with criminal justice. Thus, when the youth services division of an English town asked the simple question 'did our interventions against offending come to profit or loss last year' the answer required a considerable research effort. A criminal justice 'audit' was set in train to estimate the costs of operating the criminal justice system. In Chapter 12 Joanna Shapland contributes an account of the multidisciplinary work necessary to enable the agencies to engage in classic economic-type calculations, such as comparing the returns from work with victims to those from providing crime prevention advice. It is only by establishing upon what the criminal justice system spends its money that priorities can effectively be established. This is not just a matter for official bodies. For example, penal reform organisations and concerned citizens' groups have used such audits (Prison Reform Trust, 1998). Noting that the 'Safer Cities' programme on residential burglary reduced crime and was cost-effective, the Prison Reform Trust called for its extension. Preventing each burglary cost about £300 in high-crime areas, £900 in low-crime areas. The average cost of a burglary to the state and the victim was about £1100. From a purely economic perspective the implication was that schemes should be extended until the marginal return equalled the marginal cost. But, as the Trust pointed out, while the investment was public the return was largely private; funding would probably continue only to the point where marginal returns to the Treasury equalled the marginal cost, a point which would quickly be reached. The Trust used the explanatory apparatus of economics to anticipate likely Treasury thinking and lobby for the public interest. We can expect the criminal justice audit to be increasingly practised in future.

Traced by the chapters of this book, one can see in the development of economic analysis of crime a movement to bring a steadily widening range of factors into account. Criminology itself increasingly accepts that there is no one great 'cause' of crime. We have come to agree with

Lemert's assertion that answers to the aetiological question are likely to be multi-causal (1972). That we recognise this in respect of the bulk of criminal careers does not mean all causes are 'equal', or that we cannot recognise recurrent combinations of factors which are especially predictive. Whether or not we associate criminal offending with a wider antisocial orientation (Farrington, 1986), individuals seldom become criminal overnight in response to some key event. To put this in the context of labour market events, the individual who suffers unemployment and is involved in crime does not exhibit that behaviour simply by virtue of being unemployed. Unemployed status bears consequences other than the lack of income above welfare level. For example, those in work have open to them a range of social contacts which make of life a less isolating experience; they are subject to 'labour market discipline', which regulates their use of time; they acquire a measure of status by virtue of the work they do. None of these are directly economic factors but all are a consequence of being in work or out of it. The young man, in particular, who is out of regular employment, is denied a range of normalising relationships and experiences at a time which is, we sometimes forget, still part of the developmental socialisation cycle.

We can make a policy connection with the concept of social exclusion, an idea which has latterly preoccupied policy debates on both sides of the Atlantic (whether as debate about an 'underclass' or about 'the excluded'). Governments increasingly recognise that exclusion has many bases, not only poverty. As well as economic interventions, political, cultural and individual interventions are in frame. Examples include new approaches to the integration of migrants and immigrants into host societies which are sensitive to multiculturalism and the demise of the 'melting pot' ideology, or the mentoring programmes in youth work, where adults serve as role models and activity-organisers to help young people find a niche even when work is hard to get. None of this is to sideline the economic perspective. As we have seen, the work of economic analysis has increasing breadth, building in factors wider than the pure calculus of the pioneering economic analyses of crime and punishment. This broader perspective expands, but builds on, the economic dimensions of crime.

References

Becker, G. S. (1968) 'Crime and Punishment: an Economic Approach', *Journal of Political Economy*, 76(2), 169–217.

Farrington, D. (1986) 'Stepping Stones to Adult Criminal Careers', in D. Olweus *et al.* (eds), *Development of Antisocial and Prosocial Behaviour* (New York: Academic Press), pp. 359–84.

Fiorentini, G. and Peltzman, S. (1995) 'Introduction', in G. Fiorentini and S. Peltzman (eds), *The Economics of Organised Crime* (Cambridge: Cambridge University Press), pp. 1–30.

Gambetta, D. (1994) *The Sicilian Mafia* (Cambridge, MA: Harvard University Press).

Lemert, E. (1972) *Human Deviance, Social Problems and Social Control* (Englewood Cliffs, NJ: Prentice Hall).

Levi, M. (1998) 'Editor's Introduction to Special Issue on Organised Crime', *Howard Journal of Criminal Justice*, 37(4).

Prison Reform Trust (1998) *A Fiscal and Economic Analysis of the Crime (Sentences) Act* (London: Prison Reform Trust).

Reuter, P. (1983) *Disorganised Crime: Illegal Markets and the Mafia* (Cambridge, MA: MIT Press).

Van Duyne, P. (1993) 'Organised Crime and Business–Crime Enterprises in the Netherlands', *Crime, Law and Social Change*, 19, pp. 103–42.

—, (1996) 'The Phantom and Threat of Organised Crime', *Crime, Law and Social Change*, 24, pp. 341–77.

Wilson, J. Q. (ed.) (1983), *Crime and Public Policy* (San Francisco: ICS Press).

—, and Kelling, G. (1982) 'Broken Windows: the Police and Neighbourhood Safety', *Atlantic Monthly* (March), 127, pp. 29–38.

Part I

The Economic Analysis of Crime and Punishment

1

Editors' Introduction

When economics addresses issues of crime and punishment it works from the choices that reasoning individuals make between criminal and non-criminal courses of action, choices informed by predictions of the likely merits and demerits of available alternatives. Crimes are signs of the output of such decision-making processes. Society's effort to regulate crime registers as attempts to change elements of the mechanism of allocative choice. The costs of crime (e.g. risk of apprehension) may be increased or the benefits of crimes (e.g. the value of stolen goods) may be reduced. In this first section we focus on economic understandings of crime and punishment based on extrapolation from individual decision-making.

Economists capture these decision processes by expressing them in formal mathematical models. For those whose disciplinary background is less apt to employ formal models it is important to say two things. First, the same key postulates have been built upon by successive writers, so that a little effort to understand the models will go a long way. Second, economists are aware that the key elements of the basic models do not capture all there is to say about decisions to commit crime and/or the impact of punishment, and have sought to build in elements which capture some of the perspectives of other disciplines. There is reason for economists to feel reasonably confident about the validity of the basic assumptions of rational individuals weighing costs and benefits before choosing courses of action, but anyone developing a formal model will want it to account as fully for the phenomenon as possible. Here that means being alert to factors which may not appear instrumental, and being aware of the varying analytic purchase of models according to type of criminal behaviour.

Our first chapter undoubtedly represents a seminal contribution to the economic approach to crime and punishment. Gary Becker takes as his starting point the fact that enforcement varies greatly between different laws. Becker asks: what level of resources and how much punishment should be used to enforce different laws? These questions are addressed by formulating a measure of the social loss caused by offences and identifying punishments which minimise such loss while taking account of the costs of punishment. Becker's conception of social loss addresses criteria of vengeance, deterrence, compensation and rehabilitation. Becker's model places in relation the cost of apprehending and convicting offenders, types of punishment, and the responses offenders make to changes in enforcement.

Becker's approach gives short shrift to criminological theories of aetiology, by treating criminal behaviour as part of the more general economic theory of rational choice, where it is assumed that people will commit crimes if the expected utility exceeds the utility that could be realised by other activities. Becker's view that criminals are no different from the law-abiding in their basic motivations does not close the matter but it does allow us to construct models that test the extent to which criminal behaviour and punishment can be understood without recourse to what Becker calls the 'special' or 'ad hoc' concepts of criminology.

An example is the widespread observation that offenders seem more responsive to changes in the probability of apprehension than in the severity of punishment. Becker notes that criminology has little to say theoretically about why this should be. Becker's expected-utility approach suggests offenders are 'risk preferrers' (rather than being risk aversive). The merit of his model is that it shows why this should be so. Why offenders are 'risk preferrers' in the first place (which Becker sees as a matter of 'attitude' and ineluctable using the tools of economics) or why they gravitate to this form of risk-taking is a matter the economic model cannot resolve. But joint efforts may add value, economists showing that the utility mechanism works if offenders are seen as risk preferrers and criminologists identifying, for example, the psychology of risk preferences and the sociological processes by which such preferences are confirmed in groups organised around risk-taking.

Crime and Punishment: an Economic Approach*

Gary S. Becker

I Introduction

Since the turn of the twentieth century, legislation in Western countries has expanded rapidly to reverse the brief dominance of laissez faire during the nineteenth century. The state no longer merely protects against violations of person and property through murder, rape, or burglary but also restricts 'discrimination' against certain minorities, collusive business arrangements, 'jaywalking', travel, the materials used in construction, and thousands of other activities. The activities restricted not only are numerous but also range widely, affecting persons in very different pursuits and of diverse social backgrounds, education levels, ages, races, etc. Moreover, the likelihood that an offender will be discovered and convicted and the nature and extent of punishments differ greatly from person to person and activity to activity. Yet, in spite of such diversity, some common properties are shared by practically all legislation, and these properties form the subject matter of this essay.

In the first place, obedience to law is not taken for granted, and public and private resources are generally spent in order both to prevent offences and to apprehend offenders. In the second place, conviction is not generally considered sufficient punishment in itself; additional and sometimes severe punishments are meted out to those convicted. What determines the amount and type of resources and punishments used to enforce a piece of legislation? In particular, why does enforcement differ so greatly among different kinds of legislation?

The main purpose of this essay is to answer normative versions of these questions, namely, how many resources and how much punishment *should* be used to enforce different kinds of legislation? Put equivalently,

*This chapter was originally published in the *Journal of Political Economy*, 76(2) (1968), pp. 169–217, and is reproduced by kind permission of the University of Chicago Press.

although more strangely, how many offences *should* be permitted and how many offenders *should* go unpunished? The method used formulates a measure of the social loss from offences and finds those expenditures of resources and punishments that minimise this loss. The general criterion of social loss is shown to incorporate as special cases, valid under special assumptions, the criteria of vengeance, deterrence, compensation and rehabilitation that historically have figured so prominently in practice and criminological literature.

The optimal amount of enforcement is shown to depend on, among other things, the cost of catching and convicting offenders, the nature of punishments – for example, whether they are fines or prison terms – and the responses of offenders to changes in enforcement. The discussion, therefore, inevitably enters into issues in penology and theories of criminal behaviour. A second, although because of lack of space subsidiary, aim of this essay is to see what insights into these questions are provided by our 'economic' approach. It is suggested, for example, that a useful theory of criminal behaviour can dispense with special theories of anomie, psychological inadequacies, or inheritance of special traits and simply extend the economist's usual analysis of choice.

II Basic Analysis

(A) The Cost of Crime

Although the word 'crime' is used in the title to minimise terminological innovations, the analysis is intended to be sufficiently general to cover all violations, not just felonies – like murder, robbery and assault, which receive so much newspaper coverage – but also tax evasion, the so-called white-collar crimes, and traffic and other violations. Looked at this broadly, 'crime' is an economically important activity or 'industry', notwithstanding the almost total neglect by economists.[1] Some relevant evidence recently put together by the President's Commission on Law Enforcement and Administration of Justice (the 'Crime Commission') is reproduced in Table 1.1. Public expenditures in 1965 at the federal, state and local levels on police, criminal courts and counsel, and 'corrections' amounted to over $4 billion, while private outlays on burglar alarms, guards, counsel and some other forms of protection were about $2 billion. Unquestionably, public and especially private expenditures are significantly understated, since expenditures by many public agencies in the course of enforcing particular pieces of legislation, such as state fair-employment laws,[1] are not included, and a myriad of private precautions against crime, ranging from suburban living to taxis, are also excluded.

Table 1.1   Economic costs of crimes

Type	Costs (millions of dollars)
Crimes against persons	815
Crimes against property	3 932
Illegal goods and services	8 075
Some other crimes	2 036
Total	14 858
Public expenditures on police, prosecution, and courts	3 178
Corrections	1 034
Some private costs of combating crime	1 910
Overall total	20 980

Source: President's Commission (1967d, p. 44).

Table 1.1 also lists the Crime Commission's estimates of the direct costs of various crimes. The gross income from expenditures on various kinds of illegal consumption, including narcotics, prostitution and mainly gambling, amounted to over $8 billion. The value of crimes against property, including fraud, vandalism and theft, amounted to almost $4 billion,[3] while about $3 billion worth resulted from the loss of earnings due to homicide, assault or other crimes. All the costs listed in the table total about $21 billion, which is almost 4 per cent of reported national income in 1965. If the sizeable omissions were included, the percentage might be considerably higher.

Crime has probably become more important during the past 40 years. The Crime Commission presents no evidence on trends in costs but does present evidence suggesting that the number of major felonies per capita has grown since the early 1930s (President's Commission, 1967a, pp. 22–31). Moreover, with the large growth of tax and other legislation, tax evasion and other kinds of white-collar crime have presumably grown much more rapidly than felonies. One piece of indirect evidence on the growth of crime is the large increase in the amount of currency in circulation since 1929. For 60 years prior to that date the ratio of currency either to all money or to consumer expenditures had declined very substantially. Since then, in spite of further urbanisation and income growth and the spread of credit cards and other kinds of credit,[4] both ratios have increased sizeably.[5] This reversal can be explained by an unusual increase in illegal activity, since currency has obvious advantages over cheques in illegal transactions (the opposite is true for legal transactions) because no record of a transaction remains.[6]

(B)   The Model

It is useful in determining how to combat crime in an optimal fashion to develop a model to incorporate the behavioural relations behind the costs listed in Table 1.1. These can be divided into five categories: the relations between (1) the number of crimes, called 'offences' in this essay, and the cost of offences; (2) the number of offences and the punishments meted out; (3) the number of offences, arrests and convictions and the public expenditures on police and courts; (4) the number of convictions and the costs of imprisonments or other kinds of punishments; and (5) the number of offences and the private expenditures on protection and apprehension. The first four are discussed in turn, while the fifth is postponed until a later section.

1   Damages

Usually a belief that other members of society are harmed is the motivation behind outlawing or otherwise restricting an activity. The amount of harm would tend to increase with the activity level, as in the relation

$$H_i = H_i\,(O_i),$$

with

$$H'_i = \frac{dH_i}{dO_i} > 0,$$

(1)

where H_i is the harm from the ith activity and O_i is the activity level.[7] The concept of harm and the function relating its amount to the activity level are familiar to economists from their many discussions of activities causing external diseconomies. From this perspective, criminal activities are an important subset of the class of activities that cause diseconomies, with the level of criminal activities measured by the number of offences.

The social value of the gain to offenders presumably also tends to increase with the number of offences, as in

$$G = G(O),$$

with

$$G' = \frac{dG}{dO} > 0.$$

(2)

The net cost or damage to society is simply the difference between the harm and gain and can be written as

$$D(O) = H(O) - G(O).$$

(3)

If, as seems plausible, offenders usually eventually receive diminishing marginal gains and cause increasing marginal harm from additional offences, $G'' < 0, H'' > 0$, and

$$D'' = H'' - G'' > 0, \tag{4}$$

which is an important condition used later in the analysis of optimality positions (see, for example, the Mathematical Appendix). Since both H' and $G' > 0$, the sign of D' depends on their relative magnitudes. It follows from (4), however, that

$$D'(O) > 0 \text{ for all } O > O_a \text{ if } D'(O_a) \geq 0. \tag{5}$$

Until Section V the discussion is restricted to the region where $D' > 0$, the region providing the strongest justification for outlawing an activity. In that section the general problem of external diseconomies is reconsidered from our viewpoint, and there $D' < 0$ is also permitted.

The top part of Table 1.1 lists costs of various crimes, which have been interpreted by us as estimates of the value of resources used up in these crimes. These values are important components of, but are not identical to, the net damages to society. For example, the cost of murder is measured by the loss in earnings of victims and excludes, among other things, the value placed by society on life itself; the cost of gambling excludes both the utility to those gambling and the 'external' disutility to some clergy and others; the cost of 'transfers' like burglary and embezzlement excludes social attitudes towards forced wealth redistributions and also the effects on capital accumulation of the possibility of theft. Consequently, the $15 billion estimate for the cost of crime in Table 1.1 may be a significant understatement of the net damages to society, not only because the costs of many white-collar crimes are omitted, but also because much of the damage is omitted even for the crimes covered.

2   The Cost of Apprehension and Conviction

The more that is spent on policemen, court personnel and specialised equipment, the easier it is to discover offences and convict offenders. One can postulate a relation between the output of police and court 'activity' and various inputs of manpower, materials and capital, as in $A = f(m,r,c)$, where f is a production function summarising the 'state of the arts'. Given f and input prices, increased 'activity' would be more costly, as summarised by the relation

$$C = C(A)$$

and
(6)

$$C' = \frac{dC}{dA} > 0.$$

It would be cheaper to achieve any given level of activity the cheaper were policemen,[8] judges, counsel and juries and the more highly developed the state of the arts, as determined by technologies like fingerprinting, wire-tapping, computer control and lie-detecting.[9]

One approximation to an empirical measure of 'activity' is the number of offences cleared by conviction. It can be written as

$$A \cong pO,$$
(7)

where p, the ratio of offences cleared by convictions to all offences, is the over-all probability that an offence is cleared by conviction. By substituting (7) into (6) and differentiating, one has

$$C_p = \frac{\partial C(pO)}{\partial p} = C'O > 0$$

and
(8)

$$C_o = C'p > 0$$

if $pO \neq 0$. An increase in either the probability of conviction or the number of offences would increase total costs. If the marginal cost of increased 'activity' were rising, further implications would be that

$$C_{pp} = C''O^2 > 0,$$

$$C_{oo} = C''P^2 > 0,$$
(9)

and
$$C_{po} = C_{op} = C''pO + C' > 0.$$

A more sophisticated and realistic approach drops the implication of (7) that convictions alone measure 'activity', or even that p and O have identical elasticities, and introduces the more general relation

$$A = h(p, O, a).$$
(10)

The variable a stands for arrests and other determinants of 'activity', and there is no presumption that the elasticity of h with respect to p equals that with respect to O. Substitution yields the cost function $C = C(p, O, a)$. If, as is extremely likely, h_p, h_o, and h_a are all greater than zero, then clearly C_p, C_o, and C_a are all greater than zero.

In order to ensure that optimality positions do not lie at 'corners', it is necessary to place some restrictions on the second derivatives of the cost function. Combined with some other assumptions, it is *sufficient* that

$$C_{pp} \geq 0,$$
$$C_{oo} \geq 0, \quad (11)$$

and

$$C_{po} \cong 0$$

(see the Mathematical Appendix). The first two restrictions are rather plausible, the third much less so.[10]

Table 1.1 indicates that in 1965 public expenditures in the United States on police and courts totalled more than $3 billion, by no means a minor item. Separate estimates were prepared for each of seven major felonies.[11] Expenditures on them averaged about $500 per offence (reported) and about $2000 per person arrested, with almost $1000 being spent per murder (President's Commission, 1967a, pp. 264–65); $500 is an estimate of the average cost

$$AC = \frac{C(p, O, a)}{O}$$

of these felonies and would presumably be a larger figure if the number of either arrests or convictions were greater. Marginal costs (C_o) would be at least $500 if condition (11), $C_{oo} \geq 0$, were assumed to hold throughout.

3 The Supply of Offences

Theories about the determinants of the number of offences differ greatly, from emphasis on skull types and biological inheritance to family upbringing and disenchantment with society. Practically all the diverse theories agree, however, that when other variables are held constant, an increase in a person's probability of conviction or punishment if convicted would generally decrease, perhaps substantially, perhaps negligibly, the number of offences he commits. In addition, a common generalisation by persons with judicial experience is that a change in the probability has a greater effect on the number of offences than a change in the punishment,[12] although, as far as I can tell, none of the prominent theories sheds any light on this relation.

The approach taken here follows the economists' usual analysis of choice and assumes that a person commits an offence if the expected utility to him exceeds the utility he could get by using his time and other resources at other activities. Some persons become 'criminals', therefore, not because their basic motivation differs from that of other persons, but because their benefits and costs differ. I cannot pause to discuss the many general implications of this approach,[13] except to remark that criminal behaviour becomes part of a much more general theory and does not require ad-hoc concepts of differential association, anomie, and the like,[14] nor does it assume perfect knowledge, lightning-fast calculation or any of the other caricatures of economic theory.

This approach implies that there is a function relating the number of offences by any person to his probability of conviction, to his punishment if convicted, and to other variables, such as the income available to him in legal and other illegal activities, the frequency of nuisance arrests, and his willingness to commit an illegal act. This can be represented as

$$O_j = O_j(p_j, f_j, u_j) \tag{12}$$

where O_j is the number of offences he would commit during a particular period, p_j his probability of conviction per offence, f_j his punishment per offence, and u_j a portmanteau variable representing all these other influences.[15]

Since only convicted offenders are punished, in effect there is 'price discrimination' and uncertainty: if convicted, he pays f_j per convicted offence, while otherwise he does not. An increase in either p_j or f_j would reduce the utility expected from an offence and thus would tend to reduce the number of offences because either the probability of 'paying' the higher 'price' or the 'price' itself would increase.[16] That is,

$$O_{p_j} = \frac{\partial O_j}{\partial p_j} < 0$$

and

$$O_{f_j} = \frac{\partial O_j}{\partial f_j} < 0, \tag{13}$$

which are the generally accepted restrictions mentioned above. The effect of changes in some components of u_j could also be anticipated. For example, a rise in the income available in legal activities or an increase in law-abidingness due, say to 'education', would reduce the incentive to enter illegal activities and thus would reduce the number of offences. Or a shift in the form of the punishment, say, from a fine to imprisonment, would tend to reduce the number of offences, at least temporarily, because they cannot be committed while in prison.

This approach also has an interesting interpretation of the presumed greater response to a change in the probability than in the punishment. An increase in p_j 'compensated' by an equal percentage reduction in f_j would not change the expected income from an offence[17] but could change the expected utility, because the amount of risk would change. It is easily shown that an increase in p_j would reduce the expected utility, and thus the number of offences, more than an equal percentage increase in f_j[18] if j has preference for risk; the increase in f_j would have the greater effect if he has aversion to risk; and they would have the same effect if he is risk neutral.[19] The widespread generalisation that offenders are more

~than by the punishment when

ected-utility approach that

vant region of punishments.

all the O_j and would depend

variables are likely to differ

erences in intelligence, age,

family upbringing, etc., for

alues, p, f, and u,[20] and write

$$\tag{14}$$

kinds of properties as the

itively related to p and f and

the latter if, and only if,

Smigel (1965) and Ehrlich

n felonies reported by the

data as the basic unit of

juite stable, as evidenced by

high correlation coefficients; that there are significant negative effects on O of p and f; and that usually the effect of p exceeds that of f, indicating preference for risk in the region of observation.

A well-known result states that, in equilibrium, the real incomes of persons in risky activities are, at the margin, relatively high or low as persons are generally risk avoiders or preferrers. If offenders were risk preferrers, this implies that the real income of offenders would be lower, at the margin, than the incomes they could receive in less risky legal activities, and conversely if they were risk avoiders. Whether 'crime pays' is then an implication of the attitudes offenders have towards risk and is not directly related to the efficiency of the police or the amount spent on combating crime. If, however, risk were preferred at some values of p and f and disliked at others, public policy could influence whether 'crime pays' by its choice of p and f. Indeed, it is shown later that the social loss from illegal activities is usually minimised by selecting p and f in regions where risk is preferred, that is, in regions where 'crime does not pay'.

4   Punishments

Mankind has invented a variety of ingenious punishments to inflict on convicted offenders: death, torture, branding, fines, imprisonment, banishment, restrictions on movement and occupation, and loss of citizenship are just the more common ones. In the United States, less serious offences are punished primarily by fines, supplemented occasionally by probation, petty restrictions like temporary suspension of one's driver's licence, and imprisonment. The more serious offences are punished by a combination of probation, imprisonment, parole, fines and various restrictions on choice of occupation. A recent survey estimated for an average day in 1965 the number of persons who were either on probation, parole or institutionalised in a jail or juvenile home (President's Commission, 1967b). The total number of persons in one of these categories came to about 1 300 000, which is about 2 per cent of the labour force. About one-half were on probation, one-third were institutionalised, and the remaining one-sixth were on parole.

The cost of different punishments to an offender can be made comparable by converting them into their monetary equivalent or worth, which, of course, is directly measured only for fines. For example, the cost of an imprisonment is the discounted sum of the earnings forgone and the value placed on the restrictions in consumption and freedom. Since the earnings forgone and the value placed on prison restrictions vary from person to person, the cost even of a prison sentence of given duration is not a unique quantity but is generally greater, for example, to offenders who could earn more outside of prison.[21] The cost to each offender would be greater the longer the prison sentence, since both forgone earnings and forgone consumption are positively related to the length of sentences.

Punishments affect not only offenders but also other members of society. Aside from collection costs, fines paid by offenders are received as revenue by others. Most punishments, however, hurt other members as well as offenders: for example, imprisonment requires expenditures on guards, supervisory personnel, buildings, food, etc. Currently about $1 billion is being spent each year in the United States on probation, parole and institutionalisation alone, with the daily cost per case varying tremendously from a low of $0.38 for adults on probation to a high of $11.00 for juveniles in detention institutions (President's Commission, 1967b, pp. 193–4).

The total social cost of punishments is the cost to offenders plus the cost or minus the gain to others. Fines produce a gain to the latter that equals the cost to offenders, aside from collection costs, and so the social cost of fines is about zero, as befits a transfer payment. The social cost of probation, imprisonment and other punishments, however, generally exceeds that to offenders, because others are also hurt. The derivation of optimality conditions in the next section is made more convenient if social costs are written in terms of offender costs as

$$f' \equiv bf, \qquad (15)$$

where f' is the social cost and b is a coefficient that transforms f into f'. The size of b varies greatly between different kinds of punishments: $b \cong 0$ for fines, while $b > 1$ for torture, probation, parole, imprisonment and most other punishments. It is especially large for juveniles in detention homes or for adults in prisons, and is rather close to unity for torture or for adults on parole.

III Optimality Conditions

The relevant parameters and behavioural functions have been introduced, and the stage is set for a discussion of social policy. If the aim simply were deterrence, the probability of conviction, p, could be raised close to 1, and punishments, f, could be made to exceed the gain: in this way the number of offences, O, could be reduced almost at will. However, an increase in p increases the social cost of offences through its effect on the cost of combating offences, C, as does an increase in f if $b > 0$ through the effect on the cost of punishments, bf. At relatively modest values of p and f these effects might outweigh the social gain from increased deterrence. Similarly, if the aim simply were to make 'the punishment fit the crime', p could be set close to 1, and f could be equated to the harm imposed on the rest of society. Again, however, such a policy ignores the social cost of increases in p and f.

What is needed is a criterion that goes beyond catchy phrases and gives due weight to the damages from offences, the costs of apprehending and convicting offenders and the social cost of punishments. The social-welfare function of modern welfare economics is such a criterion, and one might assume that society has a function that measures the social loss from offences. If

$$L = L(D, C, bf, O) \qquad (16)$$

is the function measuring social loss, with presumably

$$\frac{\partial L}{\partial D} > 0, \quad \frac{\partial L}{\partial C} > 0, \quad \frac{\partial L}{\partial bf} > 0, \qquad (17)$$

the aim would be to select values of f, C, and possibly b that minimize L.

It is more convenient and transparent, however, to develop the discussion at this point in terms of a less general formulation, namely, to assume that the loss function is identical with the total social loss in real income from offences, convictions and punishments, as in

$$L = D(O) + C(p, O) + bpfO. \tag{18}$$

The term *bpfO* is the total social loss from punishments, since *bf* is the loss per offence punished and *pO* is the number of offences punished (if there is a fairly large number of independent offences). The variables directly subject to social control are the amounts spent in combating offences, *C*; the punishment per offence for those convicted, *f*; and the form of punishments, summarized by *b*. Once chosen, these variables, via the *D*, *C*, and *O* functions, indirectly determine *p*, *O*, *D*, and ultimately the loss *L*.

Analytical convenience suggests that *p* rather than *C* be considered a decision variable. Also, the coefficient *b* is assumed in this section to be a given constant greater than zero. Then *p* and *f* are the only decision variables, and their optimal values are found by differentiating *L* to find the two first-order optimality conditions,[22]

$$\frac{\partial L}{\partial f} = D'O_f + C'O_f + bpfO_f + bpO = 0. \tag{19}$$

and

$$\frac{\partial L}{\partial p} = D'O_p + C'O_p + C_p + bpfO_p + bfO = 0. \tag{20}$$

If O_f and O_p are not equal to zero, one can divide through by them, and recombine terms, to get the more interesting expressions

$$D' + C' = -bpf\left(1 - \frac{1}{\varepsilon_f}\right) \tag{21}$$

and

$$D' + C' + C_p\frac{1}{O_p} = -bpf\left(1 - \frac{1}{\varepsilon_p}\right), \tag{22}$$

where

$$\varepsilon_f = -\frac{f}{O}O_f$$

and $\tag{23}$

$$\varepsilon_p = -\frac{p}{O}O_p.$$

The term on the left side of each equation gives the marginal cost of increasing the number of offences, *O*: in equation (21) through a

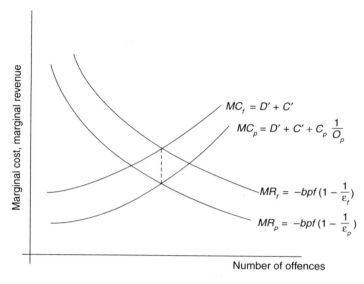

Figure 1.1 Marginal cost and marginal revenue by number of offences

reduction in f and in (22) through a reduction in p. Since $C' > 0$ and O is assumed to be in a region where $D' > 0$, the marginal cost of increasing O through f must be positive. A reduction in p partly reduces the cost of combating offences, and, therefore, the marginal cost of increasing O must be less when p rather than when f is reduced (see Fig. 1.1); the former could even be negative if C_p were sufficiently large. Average 'revenue', given by $-bpf$, is negative, but marginal revenue, given by the right-hand side of equations (21) and (22), is not necessarily negative and would be positive if the elasticities ε_p and ε_f were less than unity. Since the loss is minimized when marginal revenue equals marginal cost (see Fig. 1.1), the optimal value of ε_f must be less than unity, and that of ε_p could exceed unity only if C_p were sufficiently large. This is a reversal of the usual equilibrium condition for an income-maximising firm, which is that the elasticity of demand must exceed unity, because in the usual case average revenue is assumed to be positive.[23]

Since the marginal cost of changing O through a change in p is less than that of changing O through f, the equilibrium marginal revenue from p must also be less than that from f. But equations (21) and (22) indicate that the marginal revenue from p can be less if, and only if, $\varepsilon_p > \varepsilon_f$. As pointed out earlier, however, this is precisely the condition indicating that offenders have preference for risk and thus that 'crime does not pay'. Consequently, the loss from offences is minimised if p and f are selected from those regions where offenders are, on balance, risk preferrers. Although only the attitudes offenders have towards risk can directly determine whether 'crime pays', rational public policy indirectly insures that 'crime does not pay' through its choice of p and f.[24]

I indicated earlier that the actual p's and f's for major felonies in the United States generally seem to be in regions where the effect (measured by elasticity) of p on offences exceeds that of f; that is, where offenders are risk preferrers and 'crime does not pay' (Smigel, 1965; Ehrlich, 1967). Moreover, both elasticities are generally less than unity. In both respects, therefore, actual public policy is consistent with the implications of the optimality analysis.

If the supply of offences depended only on pf – offenders were risk neutral – a reduction in p 'compensated' by an equal percentage increase in f would leave unchanged pf, O, $D(O)$, and $bpfO$ but would reduce the loss, because the costs of apprehension and conviction would be lowered by the reduction in p. The loss would be minimised, therefore, by lowering p arbitrarily close to zero and raising f sufficiently high so that the product pf would induce the optimal number of offences.[25] A fortiori, if offenders were risk avoiders, the loss would be minimised by setting p arbitrarily close to zero, for a 'compensated' reduction in p reduces not only C but also O and thus D and $bpfO$.[26]

There was a tendency during the eighteenth and nineteenth centuries in Anglo-Saxon countries, and even today in many Communist and underdeveloped countries, to punish those convicted of criminal offences rather severely, at the same time that the probability of capture and conviction was set at rather low values.[27] A promising explanation of this tendency is that an increased probability of conviction obviously absorbs public and private resources in the form of more policemen, judges, juries and so forth. Consequently, a 'compensated' reduction in this probability obviously reduces expenditures on combating crime, and, since the expected punishment is unchanged, there is no 'obvious' offsetting increase in either the amount of damages or the cost of punishments. The result can easily be continuous political pressure to keep police and other expenditures relatively low and to compensate by meting out strong punishments to those convicted.

Of course, if offenders are risk preferrers, the loss in income from offences is generally minimised by selecting positive and finite values of p and f, even though there is no 'obvious' offset to a compensated reduction in p. One possible offset already hinted at in note 27 is that judges or juries may be unwilling to convict offenders if punishments are set very high. Formally, this means that the cost of apprehension and conviction, C, would depend not only on p and O but also on f.[28] If C were more responsive to f than p, at least in some regions,[29] the loss in income could be minimised at finite values of p and f even if offenders were risk avoiders. For then a compensated reduction in p could raise, rather than lower, C and thus contribute to an increase in the loss.

Risk avoidance might also be consistent with optimal behaviour if the loss function were not simply equal to the reduction in income. For example, suppose that the loss were increased by an increase in the ex-post 'price discrimination' between offences that are not and those that are cleared by punishment. Then a 'compensated' reduction in p would increase the 'price discrimination', and the increased loss from this could more than offset the reduction in C, D, and $bpfO$.[30]

IV Shifts in the Behavioural Relations

This section analyses the effects of shifts in the basic behavioural relations – the damage, cost and supply-of-offences functions – on the optimal values of p and f. Since rigorous proofs can be found in the Mathematical Appendix, here the implications are stressed, and only intuitive proofs are given. The results are used to explain, among other things, why more damaging offences are punished more severely and more impulsive offenders less severely.

An increase in the marginal damages from a given number of offences, D', increases the marginal cost of changing offences by a change in either p or f (see Fig. 1.2a and b). The optimal number of offences would necessarily decrease, because the optimal values of both p and f would increase. In this case (and, as shortly seen, in several others), the optimal values of p and f move in the same, rather than in opposite, directions.[31]

An interesting application of these conclusions is to different kinds of offences. Although there are few objective measures of the damages done by most offences, it does not take much imagination to conclude that offences like murder or rape generally do more damage than petty larceny or automobile theft. If the other components of the loss in income were the same, the optimal probability of apprehension and conviction and the punishment when convicted would be greater for the more serious offences.

Table 1.2 presents some evidence on the actual probabilities and punishments in the United States for seven felonies. The punishments are simply the average prison sentences served, while the probabilities are

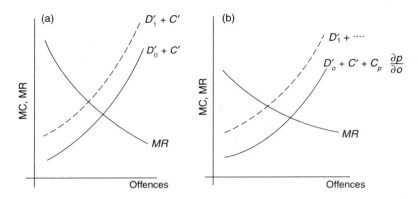

Figure 1.2   **Effect of marginal damages on marginal cost of changing offences**

ratios of the estimated number of convictions to the estimated number of offences and unquestionably contain a large error (see the discussions in Smigel, 1965, and Ehrlich, 1967). If other components of the loss function are ignored, and if actual and optimal probabilities and punishments are positively related, one should find that the more serious felonies have higher probabilities and longer prison terms. And one does: in the table, which lists the felonies in decreasing order of presumed seriousness, both the actual probabilities and the prison terms are positively related to seriousness.

Since an increase in the marginal cost of apprehension and conviction for a given number of offences, C', has identical effects as an increase in marginal damages, it must also reduce the optimal number of offences and increase the optimal values of p and f. On the other hand, an increase in the other component of the cost of apprehension and conviction, C_p, has no direct effect on the marginal cost of changing offences with f and *reduces* the cost of changing offences with p (see Fig. 1.3). It therefore reduces the optimal value of p and only partially compensates with an increase in f, so that the optimal number of offences increases. Accordingly, an increase in both C' and C_p must increase the optimal f but can either increase or decrease the optimal p and optimal number of offences, depending on the relative importance of the changes in C' and C_p.

The cost of apprehending and convicting offenders is affected by a variety of forces. An increase in the salaries of policemen increases both C' and C_p, while improved police technology in the form of fingerprinting, ballistic techniques, computer control and chemical analysis, or police

Table 1.2 Probability of conviction and average prison term for several major felonies (1960)

	Murder and non-negligent manslaughter	Forcible rape	Robbery	Aggravated assault	Burglary	Larceny	Automobile theft	All these felonies combined
1. Average time served (months) before first release:								
(a) Federal civil institutions	111.0	63.6	56.1	27.1	26.2	16.2	20.6	18.8
(b) State institutions	121.4	44.8	42.4	25.0	24.6	19.8	21.3	28.4
2. Probabilities of apprehension and conviction (percentage):								
(a) Those found guilty of offences known	57.9	37.7	25.1	27.3	13.0	10.7	13.7	15.1
(b) Those found guilty of offences charged	40.7	26.9	17.8	16.1	10.2	9.8	11.5	15.0
(c) Those entering federal and state prisons (excludes many juveniles)	39.8	22.7	8.4	3.0	2.4	2.2	2.1	2.8

Sources: 1, Bureau of Prisons (1960, Table 3); 2 (a) and (b), Federal Bureau of Investigation (1960, Table 10); 2(c), Federal Bureau of Investigation (1961, Table 2), Bureau of Prisons (n.d., Table A1; 1961, Table 8).

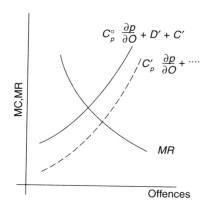

Figure 1.3   Marginal costs of apprehension and convictions

and court 'reform' with an emphasis on professionalism and merit, would tend to reduce both, not necessarily by the same extent. Our analysis implies, therefore, that although an improvement in technology and reform may or may not increase the optimal p and reduce the optimal number of offences, it does reduce the optimal f and thus the need to rely on severe punishments for those convicted. Possibly this explains why the secular improvement in police technology and reform has gone hand in hand with a secular decline in punishments.

C_p, and to a lesser extent C', differ significantly between different kinds of offences. It is easier, for example, to solve a rape or armed robbery than a burglary or automobile theft, because the evidence of personal identification is often available in the former and not in the latter offences.[32] This might tempt one to argue that the p's decline significantly as one moves across Table 1.2 (left to right) primarily because the C_p's are significantly lower for the 'personal' felonies listed to the left than for the 'impersonal' felonies listed to the right. But this implies that the f's would increase as one moved across the table, which is patently false. Consequently, the positive correlation between p, f, and the severity of offences observed in the table cannot be explained by a negative correlation between C_p (or C') and severity.

If $b > 0$, a reduction in the elasticity of offences with respect to f increases the marginal revenue of changing offences by changing f (see Fig. 1.4a). The result is an increase in the optimal number of offences and a decrease in the optimal f that is partially compensated by an increase in the optimal p. Similarly, a reduction in the elasticity of offences with respect to p also

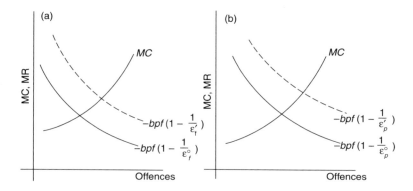

Figure 1.4 **Effect of reduced elasticity of offences**

increases the optimal number of offences (see Fig. 1.4b), decreases the optimal p, and partially compensates by an increase in f. An equal percentage reduction in both elasticities *a fortiori* increases the optimal number of offences and also tends to reduce both p and f. If $b = 0$, both marginal revenue functions lie along the horizontal axis, and changes in these elasticities have no effect on the optimal values of p and f.

The income of a firm would usually be larger if it could separate, at little cost, its total market into submarkets that have substantially different elasticities of demand: higher prices would be charged in the submarkets having lower elasticities. Similarly, if the total 'market' for offences could be separated into submarkets that differ significantly in the elasticities of supply of offences, the results above imply that if $b > 0$ the total loss would be reduced by 'charging' *lower* 'prices' – that is, lower p's and f's – in markets with *lower* elasticities.

Sometimes it is possible to separate persons committing the same offence into groups that have different responses to punishments. For example, unpremeditated murderers or robbers are supposed to act impulsively and, therefore, to be relatively unresponsive to the size of punishments; likewise, the insane or the young are probably less affected than other offenders by future consequences and, therefore,[33] probably less deterred by increases in the probability of conviction or in the punishment when convicted. The trend during the twentieth century towards relatively smaller prison terms and greater use of probation and therapy for such groups and, more generally, the trend away from the doctrine of 'a given punishment for a given crime' is apparently at least broadly consistent with the implications of the optimality analysis.

An increase in b increases the marginal revenue from changing the number of offences by changing p or f and thereby increases the optimal number of offences, reduces the optimal value of f, and increases the optimal value of p. Some evidence presented in Section II indicates that b is especially large for juveniles in detention homes or adults in prison and is small for fines or adults on parole. The analysis implies, therefore, that other things the same, the optimal f's would be smaller and the optimal p's larger if punishment were by one of the former rather than one of the latter methods.

V   Fines

(A)   Welfare Theorems and Transferable Pricing

The usual optimality conditions in welfare economics depend only on the levels and not on the slopes of marginal cost and average revenue functions, as in the well-known condition that marginal costs equal prices. The social loss from offences was explicitly introduced as an application of the approach used in welfare economics, and yet slopes as incorporated into elasticities of supply do significantly affect the optimality conditions. Why this difference? The primary explanation would appear to be that it is almost always implicitly assumed that prices paid by consumers are fully transferred to firms and governments, so that there is no social loss from payment.

If there were no social loss from punishments, as with fines, b would equal zero, and the elasticity of supply would drop out of the optimality condition given by equation (21).[34] If $b > 0$, as with imprisonment, some of the payment 'by' offenders would not be received by the rest of society, and a net social loss would result. The elasticity of the supply of offences then becomes an important determinant of the optimality conditions, because it determines the change in social costs caused by a change in punishments.

Although transferable monetary pricing is the most common kind today, the other is not unimportant, especially in underdeveloped and Communist countries. Examples in addition to imprisonment and many other punishments are the draft, payments in kind, and queues and other waiting-time forms of rationing that result from legal restrictions on pricing (see Becker, 1965) and from random variations in demand and supply conditions. It is interesting, and deserves further exploration, that the optimality conditions are so significantly affected by a change in the assumptions about the transferability of pricing.

(B) Optimality Conditions

If $b = 0$, say, because punishment was by fine, and if the cost of apprehending and convicting offenders were also zero, the two optimality conditions (21) and (22) would reduce to the same simple condition

$$D'(O) = 0. \qquad (24)$$

Economists generally conclude that activities causing 'external' harm, such as factories that pollute the air or lumber operations that strip the land, should be taxed or otherwise restricted in level until the marginal external harm equalled the marginal private gain, that is, until marginal net damages equalled zero, which is what equation (24) says. If marginal harm always exceeded marginal gain, the optimum level would be presumed to be zero, and that would also be the implication of (24) when suitable inequality conditions were brought in. In other words, if the costs of apprehending, convicting and punishing offenders were nil and if each offence caused more external harm than private gain, the social loss from offences would be minimised by setting punishments high enough to eliminate all offences. Minimising the social loss would become identical with the criterion of minimising crime by setting penalties sufficiently high.[35]

Equation (24) determines the optimal number of offences, $\hat{O}$, and the fine and probability of conviction must be set at levels that induce offenders to commit just $\hat{O}$ offences. If the economists' usual theory of choice is applied to illegal activities (see Section II), the marginal value of these penalties has to equal the marginal private gain:

$$V = G'(\hat{O}), \qquad (25)$$

where $G'(\hat{O})$ is the marginal private gain at $\hat{O}$ and V is the monetary value of the marginal penalties. Since by equations (3) and (24), $D'(\hat{O}) = H'(\hat{O}) - G'(\hat{O}) = 0$, one has by substitution in (25)

$$V = H'(\hat{O}). \qquad (26)$$

The monetary value of the penalties would equal the marginal harm caused by offences.

Since the cost of apprehension and conviction is assumed equal to zero, the probability of apprehension and conviction could be set equal to unity without cost. The monetary value of penalties would then simply equal the fines imposed, and equation (26) would become

$$f = H'(\hat{O}). \qquad (27)$$

Since fines are paid by offenders to the rest of society, a fine determined by (27) would exactly compensate the latter for the marginal harm suffered, and the criterion of minimising the social loss would be identical, at the margin, with the criterion of compensating 'victims'.[36] If the harm to victims always exceeded the gain to offenders, both criteria would reduce in turn to eliminating all offences.

If the cost of apprehension and conviction were not zero, the optimality condition would have to incorporate marginal costs as well as marginal damages and would become, if the probability of conviction were still assumed to equal unity,

$$D'(\hat{O}) + C'(\hat{O}, 1) = 0. \tag{28}$$

Since $C' > 0$, (28) requires that $D' < 0$ or that the marginal private gain exceed the marginal external harm, which generally means a smaller number of offences than when $D' = 0$.[37] It is easy to show that equation (28) would be satisfied if the fine equalled the sum of marginal harm and marginal costs:

$$f = H'(\hat{O}) + C'(\hat{O}, 1).^{[38]} \tag{29}$$

In other words, offenders have to compensate for the cost of catching them as well as for the harm they directly do, which is a natural generalisation of the usual externality analysis.

The optimality condition

$$D'(\hat{O}) + C'(\hat{O}, \hat{p}) + C_p(\hat{O}, \hat{p})\frac{1}{O_p} = 0 \tag{30}$$

would replace equation (28) if the fine rather than the probability of conviction were fixed. Equation (30) would usually imply that $D'(\hat{O}) > 0$,[39] and thus that the number of offences would exceed the optimal number when costs were zero. Whether costs of apprehension and conviction increase or decrease the optimal number of offences largely depends, therefore, on whether penalties are changed by a change in the fine or in the probability of conviction. Of course, if both are subject to control, the optimal probability of conviction would be arbitrarily close to zero, unless the social loss function differed from equation (18) (see the discussion in Section III).

(C)   The Case for Fines

Just as the probability of conviction and the severity of punishment are subject to control by society, so too is the form of punishment: legislation usually specifies whether an offence is punishable by fines, probation, institutionalisation or some combination. Is it merely an accident, or have optimality considerations determined that today, in most countries, fines are the predominant form of punishment, with institutionalisation reserved for the more serious offences? This section presents several arguments which imply that social welfare is increased if fines are used *whenever feasible*.

In the first place, probation and institutionalisation use up social resources, and fines do not, since the latter are basically just transfer payments, while the former use resources in the form of guards, supervisory personnel, probation officers and the offenders' own time.[40] Table 1.1 indicates that the cost is not minor either: in the United States in 1965, about $1 billion was spent on 'correction', and this estimate excludes, of course, the value of the loss in offenders' time.[41]

Moreover, the determination of the optimal number of offences and severity of punishments is somewhat simplified by the use of fines. A wise use of fines requires knowledge of marginal gains and harm and of marginal apprehension and conviction costs; admittedly, such knowledge is not easily acquired. A wise use of imprisonment and other punishments must know this too, however, and, in addition, must know about the elasticities of response of offences to changes in punishments. As the bitter controversies over the abolition of capital punishment suggest, it has been difficult to learn about these elasticities.

I suggested earlier that premeditation, sanity and age can enter into the determination of punishments as proxies for the elasticities of response. These characteristics may not have to be considered in levying fines, because the optimal fines, as determined, say, by equations (27) or (29), do not depend on elasticities. Perhaps this partly explains why economists discussing externalities almost never mention motivation or intent, while sociologists and lawyers discussing criminal behaviour invariably do. The former assume that punishment is by a monetary tax or fine, while the latter assume that non-monetary punishments are used.

Fines provide compensation to victims, and optimal fines at the margin fully compensate victims and restore the status quo ante, so that they are no worse off than if offences were not committed.[42] Not only do other punishments fail to compensate, but they also require 'victims' to spend additional resources in carrying out the punishment. It is not surprising, therefore, that the anger and fear felt towards ex-convicts who in fact have *not* 'paid their debt to society' have resulted in additional punishments,[43] including legal restrictions on their political and economic opportunities[44] and informal restrictions on their social acceptance. Moreover, the absence of compensation encourages efforts to change and otherwise 'rehabilitate' offenders through psychiatric counselling, therapy and other programmes. Since fines do compensate and do not create much additional cost, anger towards and fear of appropriately fined persons do not easily develop. As a result, additional punishments are not usually levied against 'ex-finees', nor are strong efforts made to 'rehabilitate' them.

One argument made against fines is that they are immoral because, in effect, they permit offences to be bought for a price in the same way that bread or other goods are bought for a price.[45] A fine *can* be considered the price of an offence, but so too can any other form of punishment; for example, the 'price' of stealing a car might be six months in jail. The only difference is in the units of measurement: fines are prices measured in monetary units, imprisonments are prices measured in time units, etc. If anything, monetary units are to be preferred here as they are generally preferred in pricing and accounting.

Optimal fines determined from equation (29) depend only on the marginal harm and cost and not at all on the economic positions of offenders. This has been criticised as unfair, and fines proportional to the incomes of offenders have been suggested.[46] If the goal is to minimise the social loss in income from offences, and not to take vengeance or to inflict harm on offenders, then fines should depend on the total harm done by offenders, and not directly on their income, race, sex, etc. In the same way, the monetary value of optimal prison sentences and other punishments depends on the harm, costs and elasticities of response, but not directly on an offender's income. Indeed, if the monetary value of the punishment by, say, imprisonment were independent of income, the length of the sentence would be *inversely* related to income, because the value placed on a given sentence is positively related to income.

We might detour briefly to point out some interesting implications for the probability of conviction of the fact that the monetary value of a given fine is obviously the same for all offenders, while the monetary equivalent or 'value' of a given prison sentence or probation period is generally positively related to an offender's income. The discussion in Section II suggested that actual probabilities of conviction are not fixed to all offenders but usually vary with their age, sex, race and, in particular, income. Offenders with higher earnings have an incentive to spend more on planning their offences, on good lawyers, on legal appeals, and even on bribery to reduce the probability of apprehension and conviction for offences punishable by, say, a given prison term, because the cost to them of conviction is relatively large compared to the cost of these expenditures. Similarly, however, poorer offenders have an incentive to use more of their time in planning their offences, in court appearances and the like to reduce the probability of conviction of offences punishable by a given fine, because the cost to them of conviction is relatively large compared to the value of their time.[47] The implication is that the probability of conviction would be systematically related to the earnings of offenders: negatively for offences punishable by imprisonment and positively for those punishable by fines. Although a negative relation for felonies and other offences punishable by imprisonment has been frequently observed and deplored (see President's Commission, 1967c, pp. 139–53), I do not know of any studies of the relation for fines or of any recognition that the observed negative relation may be more a consequence of the nature of the punishment than of the influence of wealth.

Another argument made against fines is that certain crimes, like murder or rape, are so heinous that no amount of money could compensate for the harm inflicted. This argument has obvious merit and is a special case of the more general principle that fines cannot be relied on exclusively whenever the harm exceeds the resources of offenders. For then victims could not be fully compensated by offenders, and fines would have to be supplemented with prison terms or other punishments in order to discourage offences optimally. This explains why imprisonments, proba- tion and parole are major punishments for the more serious felonies; considerable harm is inflicted, and felonious offenders lack sufficient resources to compensate. Since fines are preferable, it also suggests the need for a flexible system of instalment fines to enable offenders to pay fines more readily and thus avoid other punishments.

This analysis implies that if some offenders could pay the fine for a given offence and others could not,[48] the former should be punished solely by fine and the latter partly by other methods. In essence, therefore, these methods become a vehicle for punishing 'debtors' to society. Before the cry is raised that the system is unfair, especially to poor offenders, consider the following.

Those punished would be debtors in 'transactions' that were never agreed to by their 'creditors', not in voluntary transactions, such as loans,[49] for which suitable precautions could be taken in advance by creditors. Moreover, punishment in any economic system based on voluntary market transactions inevitably must distinguish between such 'debtors' and others. If a rich man purchases a car and a poor man steals one, the former is congratulated, while the latter is often sent to prison when apprehended. Yet the rich man's purchase is equivalent to a 'theft' subsequently compensated by a 'fine' equal to the price of the car, while the poor man, in effect, goes to prison because he cannot pay this 'fine'. Whether a punishment like imprisonment in lieu of a full fine for offenders lacking sufficient resources is 'fair' depends, of course, on the length of the prison term compared to the fine.[50] For example, a prison term of one week in lieu of a $10 000 fine would, if anything, be 'unfair' to wealthy offenders paying the fine. Since imprisonment is a more costly punishment to society than fines, the loss from offences would be reduced by a policy of leniency towards persons who are imprisoned because they cannot pay fines. Consequently, optimal prison terms for 'debtors' would not be 'unfair' to them in the sense that the monetary equivalent to them of the prison terms would be less than the value of optimal fines, which in turn would equal the harm caused or the 'debt'.[51]

It appears, however, that 'debtors' are often imprisoned at rates of exchange with fines that place a low value on time in prison. Although I have not seen systematic evidence on the different punishments actually offered convicted offenders, and the choices they made, many statutes in the United States do permit fines and imprisonment that place a low value on time in prison. For example, in New York State, Class A Misdemeanors can be punished by a prison term as long as one year or a fine no larger than $1000 and Class B Misdemeanors by a term as long as three months or a fine no larger than $500 (*Laws of New York*, 1965, chap. 1030, Arts. 70 and 80).[52] According to my analysis these statutes permit excessive prison sentences relative to the fines, which may explain why imprisonment in lieu of fines is considered unfair to poor offenders, who often must 'choose' the prison alternative.

(D) Compensation and the Criminal Law

Actual criminal proceedings in the United States appear to seek a mixture of deterrence, compensation and vengeance. I have already indicated that these goals are somewhat contradictory and cannot generally be simultaneously achieved; for example, if punishment were by fine, minimising the social loss from offences would be equivalent to compensating 'victims' fully, and deterrence or vengeance could only be partially pursued. Therefore, if the case for fines were accepted, and punishment by optimal fines became the norm, the traditional approach to criminal law would have to be significantly modified.

First and foremost, the primary aim of all legal proceedings would become the same: not punishment or deterrence, but simply the assessment of the 'harm' done by defendants. Much of traditional criminal law would become a branch of the law of torts,[53] say 'social torts', in which the public would collectively sue for 'public' harm. A 'criminal' action would be defined fundamentally not by the nature of the action[54] but by the ability of a person to compensate for the 'harm' that he caused. Thus an action would be 'criminal' precisely because it results in uncompensated 'harm' to others. Criminal law would cover all such actions, while tort law would cover all other (civil) actions.

As a practical example of the fundamental changes that would be wrought, consider the antitrust field. Inspired in part by the economist's classic demonstration that monopolies distort the allocation of resources and reduce economic welfare, the United States has outlawed conspiracies and other constraints of trade. In practice, defendants are often simply required to cease the objectionable activity, although sometimes they are also fined, become subject to damage suits, or are jailed.

If compensation were stressed, the main purpose of legal proceedings would be to levy fines equal to[55] the harm inflicted on society by constraints of trade. There would be no point to cease and desist orders, imprisonment, ridicule, or dissolution of companies. If the economist's theory about monopoly is correct, and if optimal fines were levied, firms would automatically cease any constraints of trade, because the gain to them would be less than the harm they cause and thus less than the fines expected. On the other hand, if Schumpeter and other critics are correct, and certain constraints of trade raise the level of economic welfare, fines could fully compensate society for the harm done, and yet some constraints would not cease, because the gain to participants would exceed the harm to others.[56]

One unexpected advantage, therefore, from stressing compensation and fines rather than punishment and deterrence is that the validity of the classical position need not be judged *a priori*. If valid, compensating fines would discourage all constraints of trade and would achieve the classical aims. If not, such fines would permit the socially desirable constraints to continue and, at the same time, would compensate society for the harm done.

Of course, as participants in triple-damage suits are well aware, the harm done is not easily measured, and serious mistakes would be inevitable. However, it is also extremely difficult to measure the harm in many civil suits,[57] yet these continue to function, probably reasonably well on the whole. Moreover, as experience accumulated, the margin of error would decline, and rules of thumb would develop. Finally, one must realise that difficult judgements are also required by the present antitrust policy, such as deciding that certain industries are 'workably' competitive or that certain mergers reduce competition. An emphasis on fines and compensation would at least help avoid irrelevant issues by focusing attention on the information most needed for intelligent social policy.

VI   Private Expenditures against Crime

A variety of private as well as public actions also attempt to reduce the number and incidence of crimes: guards, doormen, and accountants are employed, locks and alarms installed, insurance coverage extended, parks and neighbourhoods avoided, taxis used in place of walking or subways, and so on. Table 1.1 lists close to $2 billion of such expenditures in 1965, and this undoubtedly is a gross underestimate of the total. The need for private action is especially great in highly interdependent modern economies, where frequently a person must trust his resources, including his person, to the 'care' of employees, employers, customers, or sellers.

If each person tries to minimise his expected loss in income from crimes, optimal private decisions can be easily derived from the previous discussion of optimal public ones. For each person there is a loss function similar to that given by equation (18):

$$L_j = H_j(O_j) + C_j(p_j, O_j, C, C_k) + b_j p_j f_j O_j. \tag{31}$$

The term H_j represents the harm to j from the O_j offences committed against j, while C_j represents his cost of achieving a probability of conviction of p_j for offences committed against him. Note that C_j not only is positively related to O_j but also is negatively related to C, public expenditures on crime, and to C_k, the set of private expenditures by other persons.[58]

The term $b_j p_j f_j O_j$ measures the expected[59] loss to j from punishment of offenders committing any of the O_j. Whereas most punishments result in a net loss to society as a whole, they often produce a gain for the actual victims. For example, punishment by fines given to the actual victims is just a transfer payment for society but is a clear gain to victims; similarly, punishment by imprisonment is a net loss to society but is a negligible loss to victims, since they usually pay a negligible part of imprisonment costs. This is why b_j is often less than or equal to zero, at the same time that b, the coefficient of social loss, is greater than or equal to zero.

Since b_j and f_j are determined primarily by public policy on punishments, the main decision variable directly controlled by j is p_j. If he chooses a p_j that minimises L_j, the optimality condition analogous to equation (22) is

$$H'_j + C'_j + C_{jp_j} \frac{\partial p_j}{\partial O_j} = -b_j p_j f_j \left(1 - \frac{1}{\varepsilon_{jp_j}} \right).^{60} \qquad (32)$$

The elasticity ε_{jp_j} measures the effect of a change in p_j on the number of offences committed against j. If $b_j < 0$, and if the left-hand side of equation (32), the marginal cost of changing O_j, were greater than zero, then (32) implies that $\varepsilon_{jp_j} > 1$. Since offenders can substitute among victims, ε_{jp_j} is probably much larger than ε_p, the response of the total number of offences to a change in the average probability, p. There is no inconsistency, therefore, between a requirement from the optimality condition given by (22) that $\varepsilon_p < 1$ and a requirement from (32) that $\varepsilon_{jp_j} > 1$.

VII Some Applications

(A) Optimal Benefits

Our analysis of crime is a generalisation of the economist's analysis of external harm or diseconomies. Analytically, the generalisation consists in introducing costs of apprehension and conviction, which make the probability of apprehension and conviction an important decision variable, and in treating punishment by imprisonment and other methods as well as by monetary payments. A crime is apparently not so different analytically from any other activity that produces external harm and when crimes are punishable by fines, the analytical differences virtually vanish.

Discussions of external economies or advantages are usually perfectly symmetrical to those of diseconomies, yet one searches in vain for analogues to the law of torts and criminality. Generally, compensation cannot be controlled for the external advantages as opposed to harm caused, and no public officials comparable to policemen and district attorneys apprehend and 'convict' benefactors rather than offenders. Of course, there is public interest in benefactors: medals, prizes, titles and other privileges have been awarded to military heroes, government officials, scientists, scholars, artists and businessmen by public and private bodies. Among the most famous are Nobel Prizes, Lenin Prizes, the Congressional Medal of Honor, knighthood and patent rights. But these are piecemeal efforts that touch a tiny fraction of the population and lack the guidance of any body of law that codifies and analyses different kinds of advantages.

Possibly the explanation for this lacuna is that criminal and tort law developed at the time when external harm was more common than advantages, or possibly the latter have been difficult to measure and thus considered too prone to favouritism. In any case, it is clear that the asymmetry in the law does not result from any analytical asymmetry, for a formal analysis of advantages, benefits and benefactors can be developed that is quite symmetrical to the analysis of damages, offences, and offenders. A function $A(B)$, for example, can give the net social advantages from B benefits in the same way that $D(O)$ gives the net damage from O offences. Likewise, $K(B,p_1)$ can give the cost of apprehending and rewarding benefactors, where p_1 is the probability of so doing, with K' and $K_p > 0$; $B(p_1, a, v)$ can give the supply of benefits, where a is the award per benefit and v represents other determinants, with $\partial B/\partial p_1$ and $\partial B/\partial a > 0$; and b_1 can be the fraction of a that is a net loss to society. Instead of a loss function showing the decrease in social income from offences, there can be a profit function showing the increase in income from benefits:

$$\Pi = A(B) - K(B, p_1) - b_1 p_1 a B. \tag{33}$$

If Π is maximised by choosing appropriate values of p_1 and a, the optimality conditions analogous to equations (21) and (22) are

$$A' - K' = -b_1 p_1 a \left(1 + \frac{1}{e_a}\right) \tag{34}$$

and

$$A' - K' - K_p \frac{\partial p_1}{\partial B} = b_1 p_1 a \left(1 + \frac{1}{e_p}\right), \tag{35}$$

where

$$e_a = \frac{\partial B}{\partial a} \frac{a}{B}$$

and

$$e_p = \frac{\partial B}{\partial p_1} \frac{p_1}{B}$$

are both greater than zero. The implications of these equations are related to and yet differ in some important respects from those discussed earlier for (21) and (22).

For example, if $b_1 > 0$, which means that a is not a pure transfer but costs society resources, clearly (34) and (35) imply that $e_p > e_a$, since both $K_p > 0$ and $\partial p_1/\partial B > 0$. This is analogous to the implication of (21) and

(22) that $\varepsilon_p > \varepsilon_f$, but, while the latter implies that, at the margin, offenders are risk *preferrers*, the former implies that, at the margin, benefactors are risk *avoiders*.[61] Thus, while the optimal values of p and f would be in a region where 'crime does not pay' – in the sense that the marginal income of criminals would be less than that available to them in less risky legal activities – the optimal values of p_1 and a would be where 'benefits do pay' – in the same sense that the marginal income of benefactors would exceed that available to them in less risky activities. In this sense it 'pays' to do 'good' and does not 'pay' to do 'bad'.

As an illustration of the analysis, consider the problem of rewarding inventors for their inventions The function $A(B)$ gives the total social value of B inventions, and A' gives the marginal value of an additional one. The function $K(B,p_1)$ gives the cost of finding and rewarding inventors; if a patent system is used, it measures the cost of a patent office, of preparing applications, and of the lawyers, judges and others involved in patent litigation.[62] The elasticities e_p and e_a measure the response of inventors to changes in the probability and magnitude of awards, while b_1 measures the social cost of the method used to award inventors. With a patent system, the cost consists in a less extensive use of an invention than would otherwise occur, and in any monopoly power so created.

Equations (34) and (35) imply that with any system having $b_1 > 0$, the smaller the elasticities of response of inventors, the smaller should be the probability and magnitude of awards. (The value of a patent can be changed, for example, by changing its life.) This shows the relevance of the controversy between those who maintain that most inventions stem from a basic desire 'to know' and those who maintain that most stem from the prospects of financial awards, especially today with the emphasis on systematic investment in research and development. The former quite consistently usually advocate a weak patent system, while the latter equally consistently advocate its strengthening.

Even if A', the marginal value of an invention, were 'sizeable', the optimal decision would be to abolish property rights in an invention, that is, to set $p_1 = 0$, if b_1 and K[63] were sufficiently large and/or the elasticities e_p and e_a sufficiently small. Indeed, practically all arguments to eliminate or greatly alter the patent system have been based either on its alleged costliness, large K or b_1, or lack of effectiveness, low e_p or e_a (see, for example, Plant 1934, or Arrow, 1962).

If a patent system were replaced by a system of cash prizes, the elasticities of response would become irrelevant for the determination of optimal policies, because b_1 would then be approximately zero.[64] A system of prizes would, moreover, have many of the same other advantages that fines have in punishing offenders (see the discussion in Section V). One significant advantage of a patent system, however, is that it automatically 'meters' A', that is, provides an award that is automatically positively related to A', while a system of prizes (or of fines and imprisonment) has to estimate A' (or D') independently and often somewhat arbitrarily.

(B)   The Effectiveness of Public Policy

The anticipation of conviction and punishment reduces the loss from offences and thus increases social welfare by discouraging some offenders. What determines the increase in welfare, that is 'effectiveness', of public efforts to discourage offences? The model developed in Section III can be used to answer this question if social welfare is measured by income and if 'effectiveness' is defined as a ratio of the maximum feasible increase in income to the increase if all offences causing net damages were abolished by fiat. The maximum feasible increase is achieved by choosing optimal values of the probability of apprehension and conviction, p, and the size of punishments, f (assuming that the coefficient of social loss from punishment, b, is given).[65]

Effectiveness so defined can vary between zero and unity and depends essentially on two behavioural relations: the costs of apprehension and conviction and the elasticities of response of offences to changes in p and f. The smaller these costs or the greater these elasticities, the smaller the cost of achieving any given reduction in offences and thus the greater the effectiveness. The elasticities may well differ considerably among different kinds of offences. For example, crimes of passion, like murder or rape, or crimes of youth, like automobile theft, are often said to be less responsive to changes in p and f than are more calculating crimes by adults, like embezzlement, antitrust violation or bank robbery. The elasticities estimated by Smigel (1965) and Ehrlich (1967) for seven major felonies do differ considerably but are not clearly smaller for murder, rape, automobile theft and assault than for robbery, burglary and larceny.[66]

Probably effectiveness differs among offences more because of differences in the costs of apprehension and conviction than in the elasticities of response. An important determinant of these costs, and one that varies greatly, is the time between commission and detection of an offence.[67] For the earlier an offence is detected, the earlier the police can be brought in and the more likely that the victim is able personally to identify the offender. This suggests that effectiveness is greater for robbery than for a related felony like burglary, or for minimum-wage and fair-employment legislation than for other white-collar legislation like antitrust and public-utility regulation.[68]

(C)   A Theory of Collusion

The theory developed in this essay can be applied to any effort to preclude certain kinds of behaviour, regardless of whether the behaviour is 'unlawful'. As an example, consider efforts by competing firms to collude in order to obtain monopoly profits. Economists lack a satisfactory theory of the determinants of price and output policies by firms in an industry, a theory that could predict under what conditions perfectly competitive, monopolistic, or various intermediate kinds of behaviour would emerge. One by-product of our approach to crime and punishment is a theory of collusion that appears to fill a good part of this lacuna.[69]

The gain to firms from colluding is positively related to the elasticity of their marginal cost curves and is inversely related to the elasticity of their collective demand curve. A firm that violates a collusive arrangement by pricing below or producing more than is specified can be said to commit an 'offense' against the collusion. The resulting harm to the collusion would depend on the number of violations and on the elasticities of demand and marginal cost curves, since the gain from colluding depends on these elasticities.

If violations could be eliminated without cost, the optimal solution would obviously be to eliminate all of them and to engage in pure monopoly pricing. In general, however, as with other kinds of offences, there are two costs of eliminating violations. There is first of all the cost of discovering violations and of 'apprehending' violators. This cost is greater the greater the desired probability of detection and the greater the number of violations. Other things the same, the latter is usually positively related to the number of firms in an industry, which partly explains why economists typically relate monopoly power to concentration. The cost of achieving a given probability of detection also depends on the number of firms, on the number of customers, on the stability of customer buying patterns, and on government policies towards collusive arrangements (see Stigler, 1964).

Second, there is the cost to the collusion of punishing violators. The most favourable situation is one in which fines could be levied against violators and collected by the collusion. If fines and other legal recourse are ruled out, methods like predatory price-cutting or violence have to be used, and they hurt the collusion as well as violators.

Firms in a collusion are assumed to choose probabilities of detection, punishments to violators, and prices and outputs that minimise their loss from violations, which would at the same time maximise their gain from colluding. Optimal prices and outputs would be closer to the competitive position the more elastic demand curves were, the greater the number of sellers and buyers, the less transferable punishments were, and the more hostile to collusion governments were. Note that misallocation of resources could not be measured simply by the deviation of actual from competitive outputs but would depend also on the cost of enforcing collusions. Note further, and more importantly, that this theory, unlike most theories of pricing, provides for continuous variation, from purely competitive through intermediate situations to purely monopolistic pricing. These situations differ primarily because of differences in the 'optimal' number of violations, which in turn are related to differences in the elasticities, concentrations, legislation, etc., already mentioned.

These ideas appear to be helpful in understanding the relative success of collusions in illegal industries themselves! Just as firms in legal industries have an incentive to collude to raise prices and profits, so too do firms producing illegal products, such as narcotics, gambling, prostitution and abortion. The 'syndicate' is an example of a presumably highly successful collusion that covers several illegal products.[70] In a country like the United States that prohibits collusions, those in illegal industries would seem to have an advantage, because force and other illegal methods could be used against violators without the latter having much legal recourse. On the other hand, in countries like prewar Germany that legalised collusions, those in legal industries would have an advantage, because violators could often be legally prosecuted. One would predict, therefore, from this consideration alone, relatively more successful collusions in illegal industries in the United States, and in legal ones in prewar Germany.

VIII   Summary and Concluding Remarks

This essay uses economic analysis to develop optimal public and private policies to combat illegal behaviour. The public's decision variables are its expenditures on police, courts, etc., which help determine the probability (p) that an offence is discovered and the offender apprehended and convicted, the size of the punishment for those convicted (f), and the form of the punishment: imprisonment, probation, fine, etc. Optimal values of these variables can be chosen subject to, among other things, the constraints imposed by three behavioural relations. One shows the damages caused by a given number of illegal actions, called offences (O), another the cost of achieving a given p, and the third the effect of changes in p and f on O.

'Optimal' decisions are interpreted to mean decisions that minimise the social loss in income from offences. This loss is the sum of damages, costs of apprehension and conviction, and costs of carrying out the punishments imposed, and can be minimised simultaneously with respect to p, f, and the form of f unless one or more of these variables is constrained by 'outside' considerations. The optimality conditions derived from the minimisation have numerous interesting implications that can be illustrated by a few examples.

If carrying out the punishment were costly, as it is with probation, imprisonment, or parole, the elasticity of response of offences with respect to a change in p would generally, in equilibrium, have to exceed its response to a change in f. This implies, if entry into illegal activities can be explained by the same model of choice that economists use to explain entry into legal activities, that offenders are (at the margin) 'risk preferrers'. Consequently, illegal activities 'would not pay' (at the margin) in the sense that the real income received would be less than what could be received in less risky legal activities. The conclusion that 'crime would not pay' is an optimality condition and not an implication about the efficiency of the police or courts; indeed, it holds for any level of efficiency, as long as optimal values of p and f appropriate to each level are chosen.

If costs were the same, the optimal values of both p and f would be greater, the greater the damage caused by an offence. Therefore, offences like murder and rape should be solved more frequently and punished more severely than milder offences like automobile theft and petty larceny. Evidence on actual probabilities and punishments in the United States is strongly consistent with this implication of the optimality analysis.

Fines have several advantages over other punishments: for example, they conserve resources, compensate society as well as punish offenders, and simplify the determination of optimal p's and f's. Not surprisingly, fines are the most common punishment and have grown in importance over time. Offenders who cannot pay fines have to be punished in other ways, but the optimality analysis implies that the monetary value to them of these punishments should generally be less than the fines.

Vengeance, deterrence, safety, rehabilitation and compensation are perhaps the most important of the many desiderata proposed throughout history. Next to these, minimising the social loss in income may seem narrow, bland and even quaint. Unquestionably, the income criterion can be usefully generalised in several directions, and a few have already been suggested in the essay. Yet one should not lose sight of the fact that it is more general and powerful than it may seem and actually includes more dramatic desiderata as special cases. For example, if punishment were by an optimal fine, minimising the loss in income would be equivalent to compensating 'victims' fully and would eliminate the 'alarm' that so worried Bentham; or it would be equivalent to deterring all offences causing great damage if the cost of apprehending, convicting and punishing these offenders were relatively small. Since the same could also be demonstrated for vengeance or rehabilitation, the moral should be clear: minimising the loss in income is actually very general and thus is *more useful* than these catchy and dramatic but inflexible desiderata.

This essay concentrates almost entirely on determining optimal policies to combat illegal behaviour and pays little attention to actual policies. The small amount of evidence on actual policies that I have examined certainly suggests a positive correspondence with optimal policies. For example, it is found for seven major felonies in the United States that more damaging ones are penalised more severely, that the elasticity of response of offences to changes in p exceeds the response to f, and that both are usually less than unity, all as predicted by the optimality analysis. There are, however, some discrepancies too: for example, the actual tradeoff between imprisonment and fines in different statutes is frequently less, rather than the predicted more, favourable to those imprisoned. Although many more studies of actual policies are needed, they are seriously hampered on the empirical side by grave limitations in the quantity and quality of data on offences, convictions, costs, etc., and on the analytical side by the absence of a reliable theory of political decision-making.

Reasonable men will often differ on the amount of damages or benefits caused by different activities. To some, any wage rates set by competitive labour markets are permissible, while to others, rates below a certain minimum are violations of basic rights; to some, gambling, prostitution and even abortion should be freely available to anyone willing to pay the market price, while to others, gambling is sinful and abortion is murder. These differences are basic to the development and implementation of public policy but have been excluded from my inquiry. I assume consensus on damages and benefits and simply try to work out rules for an optimal implementation of this consensus.

The main contribution of this essay, as I see it, is to demonstrate that optimal policies to combat illegal behaviour are part of an optimal allocation of resources. Since economics has been developed to handle resource allocation, an 'economic' framework becomes applicable to, and helps enrich, the analysis of illegal behaviour. At the same time, certain unique aspects of the latter enrich economic analysis: some punishments, such as imprisonments, are necessarily non-monetary and are a cost to society as well as to offenders; the degree of uncertainty is a decision variable that enters both the revenue and cost functions etc.

Lest the reader be repelled by the apparent novelty of an 'economic' framework for illegal behaviour, let him recall that two important contributors to criminology during the eighteenth and nineteenth centuries, Beccaria and Bentham, explicitly applied an economic calculus. Unfortunately, such an approach has lost favour during the past hundred years, and my efforts can be viewed as a resurrection, modernisation and thereby I hope improvement on these much earlier pioneering studies.

Mathematical Appendix

This Appendix derives the effects of changes in various parameters on the optimal values of p and f. It is assumed throughout that $b > 0$ and that equilibrium occurs where

$$\frac{\partial D}{\partial O} + \frac{\partial C}{\partial O} + \frac{\partial C}{\partial p}\frac{\partial p}{\partial O} = D' + C' + C_p\frac{\partial p}{\partial O} > 0;$$

the analysis could easily be extended to cover negative values of b and of this marginal cost term. The conclusion in the text (Section II) that $D'' + C'' > 0$ is relied on here. I take it to be a reasonable first approximation that the elasticities of O with respect to p or f are constant. At several places a sufficient condition for the conclusions reached is that

$$C_{pO} = C_{Op} = \frac{\partial^2 C}{\partial p \partial O} = \frac{\partial^2 C}{\partial O \partial p}$$

is 'small' relative to some other terms. This condition is utilised in the form of a strong assumption that $C_{pO} = 0$, although I cannot claim any supporting intuitive or other evidence.

The social loss in income from offences has been defined as

$$L = D(O) + C(O, p) + bpfO. \tag{A1}$$

If b and p were fixed, the value of f that minimized L would be found from the necessary condition

$$\frac{\partial L}{\partial f} = 0 = (D' + C')\frac{\partial O}{\partial f} + bpf(1 - E_f)\frac{\partial O}{\partial f}, \tag{A2}$$

or

$$0 = D' + C' + bpf(1 - E_f), \tag{A3}$$

if

$$\frac{\partial O}{\partial f} = O_f \neq 0,$$

where

$$E_f = \frac{-\partial f}{\partial O}\frac{O}{f}.$$

The sufficient condition would be that $\partial^2 L/\partial f^2 > 0$; using $\partial L/\partial f = 0$ and E_f is constant, this condition becomes

$$\frac{\partial^2 L}{\partial f^2} = (D'' + C'')O_f^2 + bp(1 - E_f)O_f > 0, \qquad (A4)$$

or

$$\Delta \equiv D'' + C'' + bp(1 - E_f)\frac{1}{Of} > 0. \qquad (A5)$$

Since $D' + C' > 0$, and b is not less than zero, equation (A3) implies that $E_f > 1$. Therefore Δ would be greater than zero, since we are assuming that $D'' + C'' > 0$; and $\hat{f}$, the value of f satisfying (A3), would minimise (locally) the loss L.

Suppose that D' is positively related to an exogenous variable α. The effect of a change in α on $\hat{f}$ can be found by differentiating equation (A3):

$$D'_a + (D'' + C'')O_f\frac{d\hat{f}}{d\alpha} + bp(1 - E_f)\frac{d\hat{f}}{d\alpha} = 0,$$

or

$$\frac{d\hat{f}}{d\alpha} = \frac{-D'_\alpha(1/O_f)}{\Delta}. \qquad (A6)$$

Since $\Delta > 0$, $O_f < 0$, and by assumption $D'_\alpha > 0$, then

$$\frac{d\hat{f}}{d\alpha} = \frac{+}{+} > 0. \qquad (A7)$$

In a similar way it can be shown that, if C' is positively related to an exogenous variable β,

$$\frac{d\hat{f}}{d\beta} = \frac{-C'_\beta(1/O_f)}{\Delta} = \frac{+}{+} > 0. \qquad (A8)$$

If b is positively related to γ, then

$$(D'' + C'')O_f \frac{d\hat{f}}{d_\gamma} + bp(1 - E_f)\frac{d\hat{f}}{d_\gamma} + pf(1 - E_f)b\gamma = 0,$$

or

$$\frac{d\hat{f}}{d\gamma} = \frac{-b_\gamma pf(1 - E_f)(1/O_f)}{\Delta}. \tag{A9}$$

Since $1 - E_f < 0$, and by assumption $b_\gamma > 0$,

$$\frac{d\hat{f}}{d\gamma} = \frac{-}{+} < 0. \tag{A10}$$

Note that since $1/E_f < 1$,

$$\frac{d(p\hat{f}O)}{d\gamma} < 0. \tag{A11}$$

If E_f is positively related to δ, then

$$\frac{d\hat{f}}{d\delta} = \frac{-E_{f\delta}bpf(1/O_f)}{\Delta} = \frac{-}{+} < 0. \tag{A12}$$

Since the elasticity of O with respect to f equals

$$\epsilon_f = -O_f \frac{f}{O} = \frac{1}{E_f},$$

by (A12), a reduction in ϵ_f would reduce $\hat{f}$.

Suppose that p is related to the exogenous variable r. Then the effect of a shift in r on $\hat{f}$ can be found from

$$(D'' + C'')O_f \frac{d\hat{f}}{d_r} + (D'' + C'')O_p p_r + C_{po}p_r$$

$$+ bp(1 - E_f)\frac{\partial \hat{f}}{\partial_r} + bf(1 - E_f)p_r = 0,$$

or

$$\frac{d\hat{f}}{d_r} = \frac{-(D'' + C'')O_p(1/O_f)p_r - bf(1 - E_f)p_r(1/O_f)}{\Delta}, \tag{A13}$$

since by assumption $C_{pO} = 0$. Since $O_p < 0$, and $(D'' + C'') > 0$,

$$\frac{d\hat{f}}{d_r} = \frac{(-)+(-)}{+} = \frac{-}{+} < 0. \tag{A14}$$

If f rather than p were fixed, the value of p that minimises L, $\hat{p}$, could be found from

$$\frac{\partial L}{\partial p} = \left[D' + C' + C_p \frac{1}{O_p} + bpf(1 - E_p) \right] O_p = 0, \tag{A15}$$

as long as

$$\frac{\partial^2 L}{\partial p^2} = \left[(D'' + C'')O_p + C'_p + C_{pp}\frac{1}{O_p} + C_{pO} + C_p\frac{\partial^2 p}{\partial O\partial p} + bf(1 - E_p) \right] O_p > 0. \tag{A16}$$

Since $C'_p = C_{pO} = 0$, (A16) would hold if

$$\Delta' \equiv D'' + C'' + C_{pp}\frac{1}{O_p^2} + C_p\frac{1}{O_p}\frac{\partial^2 p}{\partial O\partial p} + bf(1 - E_p)\frac{1}{O_p} > 0. \tag{A17}$$

It is suggested in Section II that C_{pp} is generally greater than zero. If, as assumed,

$$D' + C' + C_p\frac{1}{O_p} > 0,$$

equation (A15) implies that $E_p > 1$ and thus that

$$bf(1 - E_p)\frac{1}{O_p} > 0.$$

If E_p were constant, $\partial^2 p / \partial O \partial p$ would be negative,[71] and, therefore, $C_p(1/O_p)(\partial^2 p / \partial O \partial p)$ would be positive. Hence, none of the terms of (A17) is negative, and a value of p satisfying equation (A15) would be a local minimum.

The effects of changes in different parameters on $\hat{p}$ are similar to those already derived for $\hat{f}$ and can be written without comment:

$$\frac{d\hat{p}}{d_\alpha} = \frac{-D'_\alpha(1/O_p)}{\Delta'} > 0, \tag{A18}$$

$$\frac{d\hat{p}}{d_\beta} = \frac{-C'_\beta(1/O_p)}{\Delta'} > 0, \tag{A19}$$

and

$$\frac{d\hat{p}}{d\gamma} = \frac{-b_\gamma pf(1 - E_p)(1/O_p)}{\Delta'} < 0. \tag{A20}$$

If E_p is positively related to δ',

$$\frac{d\hat{p}}{d\delta'} = \frac{E_{p\delta'} bpf(1/O_p)}{\Delta'} < 0. \tag{A21}$$

If C_p were positively related to the parameter s, the effect of a change in s on $\hat{p}$ would equal

$$\frac{d\hat{p}}{ds} = \frac{-C'_{ps}(1/O_p^2)}{\Delta'} < 0. \tag{A22}$$

If f were related to the exogenous parameter t, the effect of a change in t on $\hat{p}$ would be given by

$$\frac{d\hat{p}}{dt} = \frac{-(D'' + C'')O_p f_t(1/O_p) - bf(1 - E_p)f_t(1/O_p) - C_p(\partial^2 p/\partial O \partial f)f_t(1/O_p)}{\Delta'} < 0 \tag{A23}$$

(with $C_{pO} = 0$), since all the terms in the numerator are negative.

If both p and f were subject to control, L would be minimised by choosing optimal values of both variables simultaneously. These would be given by the solutions to the two first-order conditions, equations (A2) and (A15), assuming that certain more general second-order conditions were satisfied. The effects of changes in various parameters on these optimal values can be found by differentiating both first-order conditions and incorporating the restrictions of the second-order conditions.

The values of p and f satisfying (A2) and (A15), $\hat{p}$ and $\hat{f}$, minimise L if

$$L_{pp} > 0, L_{ff} > 0, \tag{A24}$$

and

$$L_{pp}L_{ff} > L_{fp}^2 = L_{pf}^2. \tag{A25}$$

But $L_{pp} = O_p^2 \Delta'$, and $L_{ff} = O_f^2 \Delta'$, and since both Δ' and Δ have been shown to be greater than zero, (A24) is proved already, and only (A25) remains. By differentiating L_f with respect to p and utilising the first-order condition that $L_f = 0$, one has

$$L_{fp} = O_f O_p[D'' + C'' + bf(1 - E_f)p_O] = O_f O_p \Sigma, \tag{A26}$$

where Σ equals the term in brackets. Clearly $\Sigma > 0$.

By substitution, (A25) becomes

$$\Delta\Delta' > \Sigma^2, \tag{A27}$$

and (A27) holds if Δ and Δ' are both greater than Σ. $\Delta > \Sigma$ means that

$$D'' + C'' + bp(1 - E_f)f_O > D'' + C'' + bf(1 - E_f)p_O, \tag{A28}$$

or

$$\frac{bfp}{O}(1 - E_f)E_f < \frac{bpf}{O}(1 - E_f)E_p. \tag{A29}$$

Since $1 - E_f < 0$, (A29) implies that

$$E_f > E_p, \tag{A30}$$

which necessarily holds given the assumption that $b > 0$; prove this by combining the two first-order conditions (A2) and (A15). $\Delta' > \Sigma$ means that

$$D'' + C'' + C_{pp}p_O^2 + C_p p_O p_{O_p} + bf(1 - E_p)p_O > D'' + C'' + bf(1 - E_f)p_O. \tag{A31}$$

Since $C_{pp}p_O^2 > 0$, and $p_o < 0$, this necessarily holds if

$$C_p p p_{Op} + bpf(1 - E_p) < bpf(1 - E_f). \tag{A32}$$

By eliminating $D' + C'$ from the first-order conditions (A2) and (A15) and by combining terms, one has

$$C_p p_O - bpf(E_p - E_f) = 0. \tag{A33}$$

By combining (A32) and (A33), one gets the condition

$$C_p p p_{Op} < C_p p_O, \tag{A34}$$

or

$$E_{po,p} = \frac{p}{p_O}\frac{\partial p_O}{\partial p} > 1. \tag{A35}$$

It can be shown that

$$E_{po,p} = 1 + \frac{1}{E_p} > 1, \tag{A36}$$

and, therefore, (A35) is proven.

It has now been proved that the values of p and f that satisfy the first-order conditions (A2) and (A15) do indeed minimise (locally) L. Changes in different parameters change these optimal values, and the direction and magnitude can be found from the two linear equations

$$O_f \Delta \frac{\partial \tilde{f}}{\partial z} + O_p \Sigma \frac{\partial \tilde{p}}{\partial z} = C_1$$

and

$$O_f \Sigma \frac{\partial \tilde{f}}{\partial z} + O_p \Delta' \frac{\partial \tilde{p}}{\partial z} = C_2.$$

(A37)

By Cramer's rule

$$\frac{\partial \tilde{f}}{\partial z} = \frac{C_1 O_p \Delta' - C_2 O_p \Sigma}{O_p O_f (\Delta \Delta' - \Sigma^2)} = \frac{O_p (C_1 \Delta' - C_2 \Sigma)}{+}, \quad \text{(A38)}$$

$$\frac{\partial \tilde{p}}{\partial z} = \frac{C_2 O_f \Delta - C_1 O_f \Sigma}{O_p O_f (\Delta \Delta' - \Sigma^2)} = \frac{O_f (C_2 \Delta - C_1 \Sigma)}{+}, \quad \text{(A39)}$$

and the signs of both derivatives are the same as the signs of the numerators.

Consider the effect of a change in D' resulting from a change in the parameter α. It is apparent that $C_1 = C_2 = -D'_\alpha$, and by substitution.

$$\frac{\partial \tilde{f}}{\partial \alpha} = \frac{-O_p D'_\alpha (\Delta' - \Sigma)}{+} = \frac{+}{+} > 0 \quad \text{(A40)}$$

and

$$\frac{\partial \tilde{p}}{\partial \alpha} = \frac{-O_p D'_\alpha (\Delta - \Sigma)}{+} = \frac{+}{+} > 0, \quad \text{(A41)}$$

since O_f and $O_p < 0, D'_\alpha > 0$, and Δ and $\Delta' > \Sigma$.

Similarly, if C' is changed by a change in β, $C_1 = C_2 = -C'_\beta$,

$$\frac{\partial \tilde{f}}{\partial \beta} = \frac{-O_p C'_\beta (\Delta' - \Sigma)}{+} = \frac{+}{+} > 0, \quad \text{(A42)}$$

and

$$\frac{\partial \tilde{p}}{\partial \beta} = \frac{-O_f C'_\beta (\Delta - \Sigma)}{+} = \frac{+}{+} > 0. \quad \text{(A43)}$$

If E_f is changed by a change in δ, $C_1 = E_{f\delta} bpf, C_2 = 0,$

$$\frac{\partial \tilde{f}}{\partial \delta} = \frac{O_p E_f bpf \Delta'}{+} = \frac{-}{+} < 0, \tag{A44}$$

and

$$\frac{\partial \tilde{p}}{\partial \delta} = \frac{-O_f E_f bpf \Sigma}{+} = \frac{+}{+} > 0. \tag{A45}$$

Similarly, if E_p is changed by a change in δ', $C_1 = 0$, $C_2 = E_{p\delta'} bpf$,

$$\frac{\partial \tilde{f}}{\partial \delta'} = -\frac{O_p E_{p\delta'} bpf \Sigma}{+} = \frac{+}{+} > 0, \tag{A46}$$

and

$$\frac{\partial \tilde{p}}{\partial \delta'} = \frac{O_f E_{p\delta'} bpf \Delta}{+} = \frac{-}{+} < 0. \tag{A47}$$

If b is changed by a change in γ, $C_1 = -b_\gamma pf(1 - E_f)$, $C_2 = -b_\gamma pf(1 - E_p)$, and

$$\frac{\partial \tilde{f}}{\partial \gamma} = \frac{-O_p b_\gamma pf[(1 - E_f)\Delta' - (1 - E_p)\Sigma]}{+} = \frac{-}{+} < 0, \tag{A48}$$

since $E_f > E_p > 1$ and $\Delta > \Sigma$; also,

$$\frac{\partial \tilde{p}}{\partial \gamma} = \frac{-O_f b_\gamma pf[(1 - E_p)\Delta - (1 - E_f)\Sigma]}{+} = \frac{+}{+} < 0. \tag{A49}$$

for it can be shown that $(1 - E_p)\Delta > (1 - E_f)$, Σ.[72] Note that when f is held constant the optimal value of p is decreased, not increased, by an increase in γ.

If C_p is changed by a change in s, $C_2 = -p_O C_{ps}$, $C_1 = 0$,

$$\frac{\partial \tilde{f}}{\partial s} = \frac{O_p p_O C_{ps} \Sigma}{+} = \frac{C_{ps} \Sigma}{+} = \frac{+}{+} > 0, \tag{A50}$$

and

$$\frac{\partial \tilde{p}}{\partial s} = \frac{-O_f p_O C_{ps} \Delta}{+} = \frac{-}{+} < 0. \tag{A51}$$

Notes

I would like to thank the Lilly Endowment for financing a very productive summer in 1965 at the University of California at Los Angeles. While there I received very helpful comments on an earlier draft from, among others, Armen Alchian, Roland McKean, Harold Demsetz, Jack Hirshliefer, William Meckling, Gordon Tullock and Oliver Williamson. I have also benefited from comments received at seminars at the University of Chicago, Hebrew University, RAND Corporation and several times at the Labor Workshop of Columbia; assistance and suggestions from Isaac Ehrlich and Robert Michael; and suggestions from the editor of the *Journal of Political Economy*.

1 This neglect probably resulted from an attitude that illegal activity is too immoral to merit any systematic scientific attention. The influence of moral attitudes on a scientific analysis is seen most clearly in a discussion by Alfred Marshall. After arguing that even fair gambling is an 'economic blunder' because of diminishing marginal utility, he says, 'It is true that this loss of probable happiness need not be greater than the pleasure derived from the excitement of gambling, and we are then thrown back upon the induction [*sic*] that pleasures of gambling are in Bentham's phrase "impure"; since experience shows that they are likely to engender a restless, feverish character, unsuited for steady work as well as for the higher and more solid pleasures of life' (Marshall, 1961, Note X, Mathematical Appendix).
2 Expenditures by the 13 states with such legislation in 1959 totalled almost $2 million (see Landes, 1966).
3 Superficially, frauds, thefts, etc., do not involve true social costs but are simply transfers, with the loss to victims being compensated by equal gains to criminals. While these are transfers, their market value is, nevertheless, a first approximation to the direct social cost. If the theft or fraud industry is 'competitive', the sum of the value of the criminals' time input – including the time of 'fences' and prospective time in prison – plus the value of capital input, compensation for risk, etc., would approximately equal the market value of the loss to victims. Consequently, aside from the input of intermediate products, losses can be taken as a measure of the value of the labour and capital input into these crimes, which are true social costs.
4 For an analysis of the secular decline to 1929 that stresses urbanisation and the growth in incomes, see Cagan (1965: chap. iv).
5 In 1965 the ratio of currency outstanding to consumer expenditures was 0.08, compared to only 0.05 in 1929. In 1965 currency outstanding per family was a whopping $738.
6 Cagan (1965, chap. iv) attributes much of the increase in currency holdings between 1929 and 1960 to increased tax evasion resulting from the increase in tax rates.
7 The ith subscript will be suppressed whenever it is to be understood that only one activity is being discussed.
8 According to the Crime Commission, 85–90 per cent of all police costs consist of wages and salaries (President's Commission, 1967a: 35).
9 A task-force report by the Crime Commission deals with suggestions for greater and more efficient usage of advanced technologies (President's Commission, 1967e).
10 Differentiating the cost function yields $C_{pp} = C''(h_p)^2 + C'h_{pp}$; $C_{oo} = C''(h_o)^2 + C'h_{oo}$; $C_{po} = C''h_o h_p + C'h_{po}$. If marginal costs were rising, C_{pp} or C_{oo} could be negative only if h_{pp} or h_{oo} were sufficiently negative, which is not very likely.

However, C_{po} would be approximately zero only if h_{po} were sufficiently negative, which is also unlikely. Note that if 'activity' is measured by convictions alone, $h_{pp} = h_{oo} = 0$ and $h_{po} > 0$.

11 They are wilful homicide, forcible rape, robbery, aggravated assault, burglary, larceny, and automobile theft.

12 For example, Lord Shawness (1965) said: 'Some judges preoccupy themselves with methods of punishment. This is their job. But in preventing crime it is of less significance than they like to think. Certainty of detection is far more important than severity of punishment.' Also see the discussion of the ideas of C. B. Beccaria, an insightful eighteenth-century Italian economist and criminologist, in Radzinowicz (1948: I, 282).

13 See, however, the discussions in Smigel (1965) and Ehrlich (1967).

14 For a discussion of these concepts see Sutherland (1960).

15 Both p_j and f_j might be considered distributions that depend on the judge, jury, prosecutor, etc., that j happens to receive. Among other things, u_j depends on the p's and f's meted out for other competing offences. For evidence indicating that offenders do substitute among offences, see Smigel (1965).

16 The utility expected from committing an offence is defined as

$$EU_j = p_j U_j(Y_j - f_j) + (1 - p_j)U_j(Y_j),$$

where Y_j is his income, monetary plus psychic, from an offence; U_j is his utility function; and f_j is to be interpreted as the monetary equivalent of the punishment. Then

$$\frac{\partial EU_j}{\partial p_j} = U_j(Y_j - f_j) - U_j(Y_j) < 0$$

and

$$\frac{\partial EU_j}{\partial f_j} = p_j U'_j(Y_j - f_j) < 0$$

as long as the marginal utility of income is positive. One could expand the analysis by incorporating the costs and probabilities of arrests, detentions and trials that do not result in conviction.

17 $EY_j = p_j(Y_j - f_j) + (1 - p_j)Y_j = Y_j - p_jf_j.$

18 This means that an increase in p_j 'compensated' by a reduction in f_j would reduce utility and offences.

19 From note 16

$$\frac{-\partial EU_j}{\partial p_j}\frac{p_j}{U_j} = [U_j(Y_j) - U_j(Y_j - f_j)]\frac{p_j}{U_j} \gtrless \frac{-\partial EU_j}{\partial f_j}\frac{f_j}{U_j} = p_j U'_j(Y_j - f_j)\frac{f_j}{U_j}$$

as

$$\frac{U_j(Y_j) - U_j(Y_j - f_j)}{f_j} \gtrless U'_j(Y_j - f_j).$$

The term on the left is the average change in utility between $Y_j - f_j$ and Y_j. It would be greater than, equal to, or less than $U'_j(Y_j - f_j)$ as $U''_j \gtrless 0$. But risk preference is defined by $U''_j > 0$, neutrality by $U''_j = 0$, and aversion by $U''_j < 0$.

20 p can be defined as a weighted average of the p_j, as

$$p = \sum_{j=1}^{n} \frac{O_j p_j}{\sum_{i=1}^{n} O_i}$$

and similar definitions hold for f and u.

21 In this respect, imprisonment is a special case of 'waiting time' pricing that is also exemplified by queuing (see Becker, 1965, esp. pp. 515–16, and Kleinman, 1967).

22 The Mathematical Appendix discusses second-order conditions.

23 Thus if $b < 0$, average revenue would be positive and the optimal value of ε_f would be greater than 1, and that of ε_p could be less than 1 only if C_p were sufficiently large.

24 If $b < 0$, the optimality condition is that $\varepsilon_p < \varepsilon_f$, or that offenders are risk avoiders. Optimal social policy would then be to select p and f in regions where 'crime does pay'.

25 Since $\varepsilon_f = \varepsilon_p = \varepsilon$ if O depends only on pf, and $C = 0$ if $p = 0$, the two equilibrium conditions given by equations (21) and (22)) reduce to the single condition

$$D' = -bpf\left(1 - \frac{1}{\varepsilon}\right).$$

From this condition and the relation $O = O(pf)$, the equilibrium values of O and pf could be determined.

26 If $b < 0$, the optimal solution is p about zero and f arbitrarily high if offenders are either risk neutral or risk preferrers.

27 For a discussion of English criminal law in the eighteenth and nineteenth centuries, see Radzinowicz (1948, vol. 1). Punishments were severe then, even though the death penalty, while legislated, was seldom implemented for less serious criminal offences.

Recently South Vietnam executed a prominent businessman allegedly for 'speculative' dealings in rice, while in recent years a number of persons in the Soviet Union have either been executed or given severe prison sentences for economic crimes.

28 I owe the emphasis on this point to Evsey Domar.

29 This is probably more likely for higher values of f and lower values of p.

30 If p is the probability that an offence would be cleared with the punishment f, then $1 - p$ is the probability of no punishment. The expected punishment would be $\mu = pf$, the variance $\sigma^2 = p(1 - p) f^2$, and the coefficient of variation

$$\upsilon = \frac{\sigma}{\mu} = \sqrt{\frac{1 - p}{p}};$$

υ increases monotonically from a low of zero when $p = 1$ to an infinitely high value when $p = 0$.

If the loss function equalled

$$L' = L + \psi(\upsilon), \quad \psi' > 0,$$

the optimality conditions would become

$$D' + C' = -bpf\left(1 - \frac{1}{\varepsilon_f}\right) \tag{21}$$

and

$$D' + C' + C_p\frac{1}{O_p} + \psi'\frac{dv}{dp}\frac{1}{O_p} = -bpf\left(1 - \frac{1}{\varepsilon_p}\right). \tag{22}$$

Since the term $\psi'(dv/dp)(1/O_p)$ is positive, it could more than offset the negative term $C_p(1/O_p)$.

31   I stress this primarily because of Bentham's famous and seemingly plausible dictum that 'the more deficient in certainty a punishment is, the severer it should be' (1931, chap. ii of section entitled 'Of Punishment', second rule). The dictum would be correct if p (or f) were exogenously determined and if L were minimised with respect to f (or p) alone, for then the optimal value of f (or p) would be inversely related to the given value of p (or f) (see the Mathematical Appendix). If, however, L is minimised with respect to both, then frequently they move in the same direction.

32   'If a suspect is neither known to the victim nor arrested at the scene of the crime, the chances of ever arresting him are very slim' (President's Commission, 1967e, p. 8). This conclusion is based on a study of crimes in parts of Los Angeles during January 1966.

33   But see Becker (1962) for an analysis indicating that impulsive and other 'irrational' persons may be as deterred from purchasing a commodity whose price has risen as more 'rational' persons.

34   It remains, in equation (22), through the slope O_p, because ordinarily prices do not affect marginal costs, while they do here through the influence of p on C.

35   'The evil of the punishment must be made to exceed the advantage of the offense' (Bentham, 1931, first rule).

36   By 'victims' is meant the rest of society and not just the persons actually harmed.

37   This result can also be derived as a special case of the results in the Mathematical Appendix on the effects of increases in C'.

38   Since equilibrium requires that $f = G'(\hat{O})$, and since from (28)

$$D'(\hat{O}) = H'(\hat{O}) - G'(\hat{O}) = -C'(\hat{O}, 1),$$

then (29) follows directly by substitution.

39   That is, if, as seems plausible,

$$\frac{dC}{dp} = C'\frac{\partial O}{\partial p} + C_p > 0,$$

then

$$C' + C_p\frac{1}{\partial O/\partial p} < 0,$$

and

$$D'(\hat{O}) = -\left(C' + C_p \frac{1}{\partial O/\partial p}\right) > 0.$$

40 Several early writers on criminology recognised this advantage of fines. For example, 'Pecuniary punishments are highly economical, since all the evil felt by him who pays turns into an advantage for him who receives' (Bentham, 1931, chap. vi), and 'Imprisonment would have been regarded in these old times [*ca.* tenth century] as a useless punishment; it does not satisfy revenge, it keeps the criminal idle, and do what we may, *it is costly*' (Pollock and Maitland, 1952: 516; my italics).

41 On the other hand, some transfer payments in the form of food, clothing, and shelter are included.

42 Bentham recognised this and said: 'To furnish an indemnity to the injured party is another useful quality in a punishment. It is a means of accomplishing two objects at once – punishing an offense and repairing it: removing the evil of the first order, and putting a stop to alarm. This is a characteristic advantage of pecuniary punishments' (1931: chap. vi).

43 In the same way, the guilt felt by society in using the draft, a forced transfer *to* society, has led to additional payments to veterans in the form of education benefits, bonuses, hospitalisation rights, etc.

44 See Sutherland (1960: 267–8) for a list of some of these.

45 The very early English law relied heavily on monetary fines, even for murder, and it has been said that 'every kind of blow or wound given to every kind of person had its price, and much of the jurisprudence of the time must have consisted of a knowledge of these preappointed prices' (Pollock and Maitland, 1952: 451).

The same idea was put amusingly in a recent *Mutt and Jeff* cartoon which showed a police car carrying a sign that read: 'Speed limit 30 M per H – $5 fine every mile over speed limit – pick out speed you can afford.'

46 For example, Bentham said, 'A pecuniary punishment, if the sum is fixed, is in the highest degree unequal. ... Fines have been determined without regard to the profit of the offense, to its evil, or to the wealth of the offender. ... Pecuniary punishments should always be regulated by the fortune of the offender. The relative amount of the fine should be fixed, not its absolute amount; for such an offense, such a part of the offender's fortune' (1931, chap. ix). Note that optimal fines, as determined by equation (29), do depend on 'the profit of the offence' and on 'its evil'.

47 Note that the incentive to use time to reduce the probability of a given prison sentence is unrelated to earnings, because the punishment is fixed in time, not monetary, units; likewise, the incentive to use money to reduce the probability of a given fine is also unrelated to earnings, because the punishment is fixed in monetary, not time, units.

48 In one study, about half of those convicted of misdemeanours could not pay the fines (see President's Commission, 1967c: 148).

49 The 'debtor prisons' of earlier centuries generally housed persons who could not repay loans.

50 Yet without any discussion of the actual alternatives offered, the statement is made that 'the money judgement assessed the punitive damages defendant

hardly seems comparable in effect to the criminal sanctions of death, imprisonment, and stigmatization' ('Criminal Safeguards ...', 1967)

51   A formal proof is straightforward if for simplicity the probability of conviction is taken as equal to unity. For then the sole optimality condition is

$$D' + C' = -bf\left(1 - \frac{1}{\varepsilon_f}\right). \tag{1'}$$

Since $D' = H' - G'$, by substitution one has

$$G' = H' + C' + bf\left(1 - \frac{1}{\varepsilon_f}\right), \tag{2'}$$

and since equilibrium requires that $G' = f$,

$$f = H' + C' + bf\left(1 - \frac{1}{\varepsilon_f}\right). \tag{3'}$$

or

$$f = \frac{H' + C'}{1 - b(1 - 1/\varepsilon_f)}. \tag{4'}$$

If $b > 0, \varepsilon_f < 1$ (see Section III), and hence by equation (4')

$$f < H' + C', \tag{5'}$$

where the term on the right is the full marginal harm. If p as well as f is free to vary, the analysis becomes more complicated, but the conclusion about the relative monetary value of optimal imprisonments and fines remains the same (see the Mathematical Appendix).

52   'Violations', however, can only be punished by prison terms as long as 15 days or fines no larger than $250. Since these are maximum punishments, the actual ones imposed by the courts can, and often are, considerably less. Note, too, that the courts can punish by imprisonment, by fine, or by *both* (*Laws of New York*, 1965: chap. 1030, Art. 60).

53   'The cardinal principle of damages in Anglo-American law [of torts] is that of *compensation* for the injury caused to plaintiff by defendant's breach of duty' (Harper and James, 1956: 1299).

54   Of course, many traditional criminal actions like murder or rape would still usually be criminal under this approach too.

55   Actually, fines should exceed the harm done if the probability of conviction were less than unity. The possibility of avoiding conviction is the intellectual justification for punitive, such as triple, damages against those convicted.

56   The classical view is that $D'(M)$ always is greater than zero, where M measures the different constraints of trade and D' measures the marginal damage; the critic's view is that for some M, $D'(M) < 0$. It has been shown above that if D' always is greater than zero, compensating fines would discourage all offences, in this case constraints of trade, while if D' sometimes is less than zero, some offences would remain (unless $C'[M]$, the marginal cost of detecting and convicting offenders, were sufficiently large relative to D').

57 Harper and James said: 'Sometimes [compensation] can be accomplished with a fair degree of accuracy. But obviously it cannot be done in anything but a figurative and essentially speculative way for many of the consequences of personal injury. Yet it is the aim of the law to attain at least a rough correspondence between the amount awarded as damages and the extent of the suffering' (1956: 1301).

58 An increase in $C_k - O_j$ and C held constant – presumably helps solve offences against j, because more of those against k would be solved.

59 The expected private loss, unlike the expected social loss, is apt to have considerable variance because of the small number of independent offences committed against any single person. If j were not risk neutral, therefore, L would have to be modified to include a term that depended on the distribution of $b_j p_j f_j O_j$.

60 I have assumed that

$$\frac{\partial C}{\partial p_j} = \frac{\partial C_K}{\partial p_j} = 0,$$

in other words, that j is too 'unimportant' to influence other expenditures. Although usually reasonable, this does suggest a modification to the optimality conditions given by equations (21) and (22). Since the effects of public expenditures depend on the level of private ones, and since the public is sufficiently 'important' to influence private actions, equation (22) has to be modified to

$$D' + C' + C_p \frac{\partial p}{\partial O} + \sum_{i-1}^{n} \frac{dC}{dC_i} \frac{dC_i}{dp} \frac{\partial p}{\partial O} = -bpf\left(1 + \frac{1}{\varepsilon_p}\right), \qquad (22')$$

and similarly for equation (21). 'The' probability p is, of course, a weighted average of the p_j. Equation (22') incorporates the presumption that an increase in public expenditures would be partially thwarted by an induced decrease in private ones.

61 The relation $e_p > e_a$ holds if, and only if,

$$\frac{\partial EU}{\partial p_1} \frac{p_1}{U} > \frac{\partial EU}{\partial a} \frac{a}{U}, \qquad (1')$$

where

$$EU = p_1 U(Y + a) + (1 - p_1) U(Y) \qquad (2')$$

(see the discussion on pp. 22–3). By differentiating equation (2'), one can write (1') as

$$p_1 [U(Y + a) - U(Y)] > p_1 a U'(Y + a), \qquad (3')$$

or

$$\frac{U(Y + a) - U(Y)}{a} > U'(Y + a). \qquad (4')$$

But (4′) holds if everywhere $U'' < 0$ and does not hold if everywhere $U'' \geq 0$, which was to be proved.

62   These costs are not entirely trivial: for example, in 1966 the US Patent Office alone spent $34 million (see Bureau of the Budget, 1967), and much more was probably spent in preparing applications and in litigation.

63   Presumably one reason patents are not permitted on basic research is the difficulty (that is, cost) of discovering the ownership of new concepts and theorems.

64   The right side of both (34) and (35) would vanish, and the optimality conditions would be

$$A' - K' = 0 \tag{34'}$$

and

$$A' - K' - K_p \frac{\partial p_1}{\partial B} = 0. \tag{35'}$$

Since these equations are not satisfied by any finite values of p_1 and a, there is a difficulty in allocating the incentives between p_1 and a (see the similar discussion for fines in Section V).

65   In symbols, effectiveness is defined as

$$E = \frac{D(O_1) - [D(\hat{O}) + C(\hat{p}, \hat{O}) + b\hat{p}\hat{f}\hat{O}]}{D(O_1) - D(O_2)},$$

where $\hat{p}$, $\hat{f}$ and $\hat{O}$ are optimal values, O_1 offences would occur if $p = f = 0$, and O_2 is the value of O that minimises D.

66   A theoretical argument that also casts doubt on the assertion that less 'calculating' offenders are less responsive to changes in p and f can be found in Becker (1962).

67   A study of crimes in parts of Los Angeles during January 1966 found that 'more than half the arrests were made within 8 hours of the crime, and almost two-thirds were made within the first week' (President's Commission 1967e: 8).

68   Evidence relating to the effectiveness of actual, which are not necessarily optimal, penalties for these white-collar crimes can be found in Stigler (1962, 1966), Landes (1966), and Johnson (1967).

69   Jacob Mincer first suggested this application to me.

70   An interpretation of the syndicate along these lines is also found in Schilling (1967).

71   If E_p and E_f are constants, $O = kp^{-a}f^{-b}$ where $a = 1/E_p$ and $b = 1/E_f$. Then

$$\frac{\partial p}{\partial O} = -\frac{1}{ka} p^{a+1}f^{b},$$

and

$$\frac{\partial^2 p}{\partial O \partial p} = \frac{-(a+1)}{ka} p^{a}f^{b} < 0.$$

72 The term $(1 - E_p)\Delta$ would be greater than $(1 - E_f)\Sigma$ if

$$(D'' + C'')\,(1 - E_p) + bp(1 - E_f)(1 - E_p)f_O > (D'' + C'')(1 - E_f) + bf(1 - E_f)^2 p_O,$$

or

$$(D'' + C'')\,(E_f - E_p) > -\frac{bpf}{O}(1 - E_f)\left[(1 - E_p)\frac{f_o O}{f} - (1 - E_f)\frac{p_o O}{p}\right],$$

$$(D'' + C'')\,(E_f - E_p) > -\frac{bpf}{O}(1 - E_f)[(1 - E_p)(E_f) - (1 - E_f)E_p],$$

$$(D'' + C'')\,(E_f - E_p) > -\frac{bpf}{O}(1 - E_f)(E_f - E_p).$$

Since the left-hand side is greater than zero, and the right-hand side is less than zero, the inequality must hold.

References

Arrow, Kenneth J. (1962) 'Economic Welfare and Allocation of Resources for Invention', in National Bureau Committee for Economic Research, *The Rate and Direction of Inventive Activity: Economic and Social Factors* (Princeton, NJ: Princeton University Press for the Nat. Bureau of Econ. Res.).

Becker, Gary S. (1962) 'Irrational Behavior and Economic Theory', *Journal of Political Economy*, LXX (February).

— (1965) 'A Theory of the Allocation of Time', *Economic Journal*, LXXV (September).

Bentham, Jeremy (1931) *Theory of Legislation* (New York: Harcourt Brace).

Bureau of the Budget (1967) *The Budget of United States Government, 1968, Appendix.* (Washington, DC: US Government Printing Office).

Bureau of Prisons (1960) *Prisoners Released from State and Federal Institutions* ('National Prisoner Statistics') (Washington, DC: US Dept. of Justice).

— *Characteristics of State Prisoners, 1960* ('National Prisoner Statistics') (Washington, DC: US Dept. of Justice).

— (1961) *Federal Prisons, 1960* (Washington, DC: US Dept. of Justice).

Cagan, Phillip (1965) *Determinants and Effects of Changes in the Stock of Money 1875–1960* (New York: Columbia University Press, for the Nat. Bureau of Econ. Res.).

— (1967) 'Criminal Safeguards and the Punitive Damages Defendant' (1967), *University of Chicago Law Review*, XXXIV (Winter).

Ehrlich, Isaac (1967) 'The Supply of Illegitimate Activities'. Unpublished manuscript, Columbia University, New York.

Federal Bureau of Investigation (1960) *Uniform Crime Reports for the United States* (Washington, DC: US Dept. of Justice).

— (1961) *Uniform Crime Reports for the United States* (Washington, DC: US Dept. of Justice).

Harper, F. V. and James, F. (1956) *The Law of Torts*, vol. II (Boston: Little-Brown).

Johnson, Thomas (1967) 'The Effects of the Minimum Wage Law'. Unpublished PhD dissertation, Columbia University, New York.

Kleinman, E. (1967) 'The Choice between Two "Bads" – Some Economic Aspects of Criminal Sentencing.' Unpublished manuscript, Hebrew University, Jerusalem.

Landes, William (1966) 'The Effect of State Fair Employment Legislation on the Economic Position of Nonwhite Males'. Unpublished PhD dissertation, Columbia University, New York.

Laws of New York, vol. II (1965).

Marshall, Alfred. (1961) *Principles of Economics*, 8th edn (New York: Macmillan).

Plant, A. (1934) 'The Economic Theory concerning Patents for Inventions', *Economica*, I (February).

Pollock, F. and Maitland, F. W. (1952) *The History of English Law*, vol. II, 2nd edn (Cambridge: Cambridge University Press).

President's Commission on Law Enforcement and Administration of Justice (1967a) *The Challenge of Crime in a Free Society* (Washington, DC: US Government Printing Office).

— (1967b) *Corrections* ('Task Force Reports') (Washington, DC: US Government Printing Office).

— (1967c) *The Courts* ('Task Force Reports') (Washington, DC: US Government Printing Office).

— (1967d) *Crime and its Impact – an Assessment* ('Task Force Reports') (Washington, DC: US Government Printing Office).

— (1967e) *Science and Technology* ('Task Force Reports') (Washington, DC: US Government Printing Office).

Radzinowicz, L. (1948) *A History of English Criminal Law and its Administration from 1750*, vol. I (London: Stevens & Sons).

Schilling, T. C. (1967) 'Economic Analysis of Organised Crime', in President's Commission on Law Enforcement and Administration of Justice, *Organised Crime* ('Task Force Reports') (Washington, DC: US Government Printing Office).

Shawness, Lord (1965) 'Crime Does Pay because We Do Not Back Up the Police', *New York Times Magazine*, 13 June.

Smigel, Arleen (1965) 'Crime and Punishment: an Economic Analysis'. Unpublished MA thesis, Columbia University, New York.

Stigler, George J. (1962) 'What Can Regulators Regulate? The Case of Electricity', *Journal of Law and Economics*, V (October).

— (1964) 'A Theory of Oligopoly', *Journal of Political Economy*, LXXII (February).

— (1966) 'The Economic Effects of the Antitrust Laws', *Journal of Law and Economics*, IX (October).

Sutherland, E. H. (1960) *Principles of Criminology*, 6th edn (Philadelphia: J. B. Lippincott).

2

Editors' Introduction

Antony Dnes's chapter develops an economic approach to criminal deterrence which helps us to understand why some injurious behaviours are dealt with as civil matters and others as crimes. Taking his lead from Becker's analysis of criminals as rational maximizers, Dnes explores the deterrence hypothesis that criminal activity can be reduced by increasing the costs and decreasing the benefits of crime. Dnes shows that the real question is which factors in the model significantly influence criminal behaviour. He notes that while there is support for the hypothesis, results vary for different crimes, different weights apply to the effects of detection and conviction compared to the effect of the severity of sentencing (Becker argued the former would be more efficacious than the latter), but also that conclusive testing of the hypothesis is difficult due to problems in obtaining consistent longitudinal data.

This strand of research suggests different offences display different elasticity relative to the clear-up rate (non-sexual violent crime shows little elasticity to this deterrence variable while property-related crime shows significant elasticity). Further, both probability of conviction and severity of sentence register deterrent effects, with the former stronger than the latter. However, economic analysis collides with criminal justice realities in the finding that, despite technical efficiency, the cost of increasing probability of conviction far exceeds that of more severe sentencing, making the latter the best option in practical terms.

Dnes also examines the unemployment hypothesis which competes with the deterrence hypothesis, and applies deterrence theory to the hotly contested issue of capital punishment. While some research suggests a general deterrent effect for capital punishment, Dnes maintains this does not necessarily make a strong economic argument for it.

The Economics of Crime

Antony W. Dnes

Economics examines crime as a special case of maximising behaviour. The literature has developed separately from the economic analysis of law as a part of mainstream applied economics, as economists have mostly concerned themselves, as does this chapter, with the economics of criminal deterrence. We examine deterrence hypotheses, empirical tests of deterrence, other influences on crime such as unemployment, and distinctions between violent and non-violent crime.

Criminal Activity

It is not obvious why many injurious acts are regarded as crimes when others are not. Indeed, some crimes, such as unfinished robberies, can be 'victimless' and many actions now treated as crimes were once treated as torts. What are the features of criminal law that set it apart from its civil counterpart?

The standard of proof is different in criminal cases. First, the prosecution must prove its case beyond reasonable doubt, which is a tough criterion. In a tort case it is enough to show that the defendant was negligent by the standards of a 'reasonable' person. The tougher criterion in a criminal case reflects the severe penalties the court may apply, which go beyond compensating the victim and impose harm on the criminal.

Secondly, the prosecution must usually show that the criminal intended to commit the crime. This again contrasts with civil cases where, for example, nuisance is defined independently of the state of mind of the tortfeasor. One explanation is that criminal law aims to suppress the criminal frame of mind, identified by intent. The non-contradictory exception to the requirement to show intent is the case of the strict-liability crime, such as possession of an offensive weapon, where possession may reasonably be taken to imply intent to use.

Criminal harm has an element of publicness about it as it disrupts beneficial codes of behaviour. Murder, assault and robbery are disturbing to the wider population. As Shavell (1985 and 1993) implies, the victim of a crime has an incentive to sue to reclaim personal losses only. This will not compensate the fear and anxiety experienced by dispersed observers of the crime. Bear in mind that the victim can take an action in tort against a criminal: the criminal law comes in on top of civil remedies.

There is often a low probability of apprehension in the case of criminal acts, particularly as the criminal may deliberately hide the crime (Posner, 1985) and severe penalties will be needed to deter the harm. If there is only a 5 per cent chance of catching the thief of a bracelet valued at £1000, the fine must be set at just over £20 000 to make the expected value of stealing negative. Action in tort would succeed only in reclaiming £1000 damages. Since criminals are often impoverished types who would not be able to pay damages, the only way to create a deterrent may be to impose a non-monetary sanction.

Crime is socially wasteful and represents a classic form of rent-seeking behaviour (Tullock, 1968) because criminals devote resources to pure redistribution of existing goods without creating anything new. Law-abiding citizens also buy security devices, pay insurance companies and fund the police. Finally, since the thief may value a stolen item at an amount below its owner's valuation, resources may move to lower-valued uses when items are stolen.

Criminal Deterrence

According to Becker (1968) if criminals are rational maximisers:

   (i) crime rates respond to the costs and benefits of committing crimes, and
  (ii) people respond to deterring incentives.

Devoting resources to detection, conviction and punishment should influence the level of crime. As Buchanan and Hartley (1992) point out, people need not be rational all the time. If varying the penalty influences criminal behaviour at the margin, deterrence will work.

Alternatives to the 'deterrence hypothesis' are claims that crime results from biological influences like mental illness or from social factors such as unemployment. The alternative implications are that we need to tackle unemployment and poverty and/or improve the mental health of the population. Given rising rates for property crimes in the 1990s, we would have to show that living standards had been falling or that the mental health of the criminal classes had suddenly deteriorated. Strictly biological accounts of crime are also often subject to criticism. Buchanan and Hartley (1992) point out that, if criminality were genetically determined, it would be hard to explain why the original convict population of Australia became law-abiding. Furthermore, the observation that members of the criminal classes are frequently of below average intelligence (Wilson and Herrnstein, 1985) is consistent with the deterrence hypothesis: low intelligence may limit their options, leading them rationally to turn to crime. The same could be true of mental illness or poverty.

The deterrence hypothesis suggests that individuals balance off the costs and benefits of crime in deciding whether to engage in criminal activity. For some people the balancing calculation will imply they do not engage in crime. As one example, high costs of criminal activity could be consistent with worrying about the effect on one's reputation if criminal associations were apparent. The costs of criminal activity are not necessarily just monetary ones associated with preparing to commit crimes. Similarly, benefits are not limited to monetary gains: for example, some evil and sadistic thugs obtain direct pleasure from engaging in acts of violence.

We can influence individuals to reduce their criminal activity by undertaking policies to increase the costs and decrease the benefits to them from crime. This may involve influencing criminals to reduce criminal activity, or stopping some of them from at all engaging in crime. Comparing the costs and benefits of clearing up a particular crime allows us to find an optimum level of crime prevention. It is important to remember that no policies are costless, and it is unlikely to be sensible to try to stop all crime, which would almost certainly be too costly to achieve.

The deterrence approach can be compatible with alternative approaches emphasising wider social factors or apparently non-economic individual characteristics. For example, encouraging a general revulsion towards crime might create high psychic (conscience-based) costs from engaging in crime, and reduce crime levels. The real question is empirical: which factors influence criminal behaviour in a significant way?

The deterrence hypothesis can be formulated, following Becker (1968), mathematically. Taking any crime as an example:

$$EU = pU(Y - f) + (1 - p)U(Y)$$

where: p = probability of capture and punishment
U = utility (i.e., psychic benefit, assumed measurable)

EU = expected utility
f = value of punishment
Y = income if undetected

Increasing p by a small amount (dp) implies EU changes by $dp[U(Y - f) - U(Y)]$, which must be negative. This result implies that increasing the probability of detection, perhaps by expanding the police force, deters criminal behaviour. Similarly, increasing the severity of the punishment (f) by a small amount (df) implies EU changes by $df(p)(\partial U/\partial f)$, which is again negative since $\partial U/\partial f$ (the rate of change of utility with respect to punishment) should be negative. Both increasing the probability of capture and increasing the severity of the punishment should deter crime. Becker (1968) shows that increasing the probability of capture (and punishment) should deter more strongly than increasing the severity of sentencing, providing criminals are risk takers (Dnes, 1996).

There are several reasons why deterrence might not be found (Cameron, 1988). Individuals might be less vigilant about crime if they felt the authorities had it under control. Secondly, some crime could be displaced to another offence type, time or location. Finally, the deterrence of established criminals could encourage the entry of replacements into the crime 'industry'.

Tests of the Deterrence Hypothesis

Statistical tests of the deterrence hypothesis examine the effects of a range of criminal-justice and more general variables in determining a variety of types of crime (Nagin, 1998). Property crime mainly comprises burglary and theft, whereas violent crime consists of robberies like mugging as well as the more obvious categories of murder and assault. One surprising result of the empirical work is that expenditure on the police does not appear to deter crime (Carr-Hill and Stern, 1979), which has been interpreted as indicating a failure for statistical studies to support the deterrence hypothesis (Cameron, 1988). However, expenditure on the police is not necessarily correlated with arrests if the extra members of the force are desk-bound.

Statistical studies of deterrence support the deterrence hypothesis but give different detailed results for different crimes. Secondly, they give different weights to the effect of detection and conviction relative to the severity of sentencing in deterring crime. Finally, studies can be plagued by data problems as it is difficult to obtain consistent long-run data in studies for the USA owing to the separate state jurisdictions.

Ehrlich (1973) examined US robbery data for 1940, 1950 and 1960, and used regression techniques. He found that a higher probability of conviction implied a lower robbery rate. In addition, a higher average sentence implied a lower robbery rate for 1940 and 1960. This study gave early support to the deterrence hypothesis. One criticism of it is that data on the criminal justice system and crime variables may not be fully comparable across different states. Consequently, Blumstein and Nagin (1977) examined evasion of the US military draft in the 1960s and 1970s, as these data are consistent across different states. A greater probability of conviction implied lower evasion of the draft, supporting the deterrence hypothesis.

Wolpin (1978a) examined (uniform) data from England and Wales for the period 1894–1967 using time-series analysis. He found evidence of a deterrence effect from both the probability of punishment and from the severity of sentencing, particularly for property crime. Wolpin's results show that the deterrent from increasing the probability of punishment exceeds that from increasing the severity of sentencing, consistent with Becker's (1968) suggestion.

Willis (1983) also used data from England and Wales, and used the clear-up rate for particular crimes as a measure of the probability of capturing and punishing criminals. Although Willis's focus was partly on regional variations in crime, the study usefully produced general results supporting Wolpin (1978a). Again, the strongest deterrence came from increasing the probability of punishment, although there was also deterrence from increasing the severity of sentences. Willis's study highlights differences between categories of crime. A 1 per cent increase in the clear-up rate:

(i)   reduces thefts by 0.8 per cent
(ii) reduces sex crimes by 1 per cent, and
(iii) has no effect on violent crime.

These figures describe the *elasticity* of crime with respect to the clear-up rate. This elasticity can be defined as the proportional change in the crime variable in response to a 1 per cent increase in a deterrence variable (the clear-up rate) with a higher figure indicating greater responsiveness. Non-sexual violent crime appears to be completely inelastic in this study. Violence may be less firmly based in rational behaviour than is property crime.

Work carried out by Pyle (1983) using UK data for the period 1950–80 also found deterrence from both the probability of conviction and from increases in the severity of sentences. He also found that increasing the

Table 2.1   Cost of reducing property crime by 1 per cent

Option	Cost (£m)
Increased police numbers	51.2
Increased number of offenders imprisoned	4.9
Increased average length of sentence	3.6

Source: Pyle (1989).

probability of conviction had the stronger deterrence effect. The elasticity of offences with respect to the conviction rate was 0.9, but was only 0.3 with respect to sentence length.

In a study based on data from Australian states and territories, Withers (1984) concluded that the major reliable determinants of crime rates were committal and imprisonment rates. Committal rates reflect the probability of trial, whereas the imprisonment rate is a severity measure. Despite Withers' (1984: 182) 'prior expectation', television-viewing habits were not statistically significant determinants of crime.

Finally, just because an instrument is technically highly deterring does not mean that it is good value for money: a less deterring but cheaper one might be better. Pyle (1983, 1989, 1993) has developed his work in terms of assessing the cost-effectiveness of alternative instruments of deterrence. The cost of achieving a 1 per cent reduction in property crime by alternative means is shown in Table 2.1.

Although the elasticity of deterrence is technically higher for increases in the probability of conviction, increasing the severity of sentences is more cost-effective as a means of beating property crime. Increasing police numbers to increase the probability of conviction looks like particularly poor value.

Unemployment and Crime

The effect on crime rates of changes in unemployment levels has been subjected to statistical testing. The unemployment hypothesis suggests people are driven to crime by deprivation, and is often interpreted as an alternative to the deterrence hypothesis. However, the unemployment hypothesis need not be inconsistent with Becker's (1968) view of crime: the unemployed have a lower opportunity cost attached to using their time in criminal pursuits. In fact there turns out to be little support in empirical work for an effect from unemployment.

Cook and Zarkin (1985) use regression techniques on US data to show that there are small increases in burglaries and robberies during times of recession. However, their results show there is no effect on homicide rates–and thefts of motor vehicles actually fall in a recession and rise in the up-swing. The studies by Wolpin (1978a) and Willis (1983), discussed in connection with the deterrence hypothesis, also showed a weak effect on crime from unemployment. Similar conclusions were reached by Pyle (1989) and by Withers (1984).

Field (1990) examined changes in consumption levels rather than unemployment to assess the effects of deprivation on crime in the UK. Consumption changes may have a more immediate motivational impact. He found evidence of a 'bounce-back' effect. Decreases in consumption were initially followed by an increase in property crime, which returned to normal trend value in the longer term. Field concluded that there is no real long-run relationship between changes in consumption levels and crime rates.

Using data for Scotland, Reilly and Witt (1992) estimated several models that supported a robust link between crime and unemployment. Their data comprised 15 annual observations from 1974 to 1988 for each of the six regions of Scotland, taking the general crime rate as the dependent variable. Pyle and Deadman (1994) could not replicate Reilly and Witt's results using a data set updated to include 1988–91. The earlier study does incorporate a period of falling unemployment from 1987 to 1990, which occurred while crime continued on a rising trend. Adding three observations to the data set rendered the unemployment variable statistically insignificant. Pyle and Deadman also carried out a separate time-series analysis, using quarterly data to increase the available number of observations, which also failed to find a statistically significant role for unemployment.

Pyle and Deadman (1994) point out that testing theories of crime using cross-sectional data (e.g. looking across different regions) can be problematic. A correlation between crime and unemployment, for example, may not show a causal relationship but, rather, may reflect the influence of a third variable such as poor educational standards. Also, there may be crime spillovers: an area with high unemployment may be less able to raise local taxes to spend on policing and may attract criminals from other areas. The unemployed would not be committing the crimes but researchers would find a high correlation between unemployment and crime.

Violent Crime is Different

Most countries have experienced increased levels of violent crime in recent years. In the UK, violence against the person, which includes murder, wounding and assault, grew from under 10 000 to over 200 000 annual incidents between 1946 and 1998 with most of this growth occurring in assaults. The figure for 1998 is roughly equivalent to a rate of 363 per 100 000 of population.

Violent crime may be subject to a wider range of influences than property crime. Withers (1984) found that deterrence variables were only weakly significant determinants of violent crimes. Secondly, studies of capital punishment, which we look at below, sometimes support the deterrence hypothesis in relation to murder (Ehrlich, 1975, 1977) and sometimes do not (Passell and Taylor, 1977). Results like these reveal a role for wider influences on violent crime.

Field (1990) found a positive correlation between violent crime and the level of economic activity. The explanation offered is that boom times lead to increased use of leisure facilities and greater inter-personal contact. Field also found a positive correlation between the consumption of alcohol and the incidence of violent offences. Walmsely (1986) reaches similar conclusions and cites detailed work showing that over 25 per cent of violent incidents in some British cities occur immediately after the public houses close. Policy responses to these findings could include such things as staggering closure times and increasing policing in sensitive areas at closure times, or perhaps increasing the cost of alcoholic beverages (Pyle, 1993).

However, some empirical work shows that criminal-justice variables also deter violent crime. In the USA, Murray and Cox (1979) tracked 317 youths after they were released from their first custodial sentences. The youths had an average record of 13 arrests each prior to their imprisonment for an average of 10 months. Their offences included homicide, rape, assaults, car theft, armed robbery and burglary. After imprisonment their arrest records fell by two-thirds on average. A comparison group of non-imprisoned juvenile offenders did not show the same reduction in the rate of re-arrest. Although it is possible that the imprisoned youths were learning from other crooks how to avoid arrest, such a large difference makes it unlikely: as Murray and Cox comment, no school is as good as that!

Witte (1980) examined the characteristics of 641 men released over a three-year period in North Carolina. The higher was the probability of conviction for a crime, the lower was the number of subsequent arrests per month released. Increases in the severity of prior punishment had a stronger deterrent effect compared with increases in the probability of conviction for violent crime. Increasing the probability of conviction had a stronger deterrent effect on re-offending in the case of crimes against property.

Imprisonment v. Fines

Imprisoning an offender incapacitates that person and prevents further offences being committed against the general public. At the same time, rehabilitation of the criminal is made possible. Also, we know from earlier sections of this chapter that there is a deterrent effect on the incarcerated and on others.

The costs of imprisonment are generally high, however, which explains why alternatives such as fines are often used. First there are direct costs of imprisonment. In the UK in 1998 these were an average of approximately £35 000 per year for each prisoner, although the costs of special units can be much higher at around £85 000. There are also less obvious costs in the shape of the opportunity costs of prisoners' non-criminal skills, which they cannot use while incarcerated.

Fines raise money for the state and are low-cost measures. They are a very attractive option where there are no special reasons to use imprisonment. Typically it is the limit to the criminal's wealth that causes the state to resort to non-pecuniary penalties like imprisonment. Other sanctions comprise orders for probation or for compulsory community service. In the US case, unusually among advanced societies, there is also the possibility of capital punishment.

Capital Punishment

Capital punishment is mostly abolished in developed countries, although there have been periodic debates in the UK about whether it should be reintroduced. In the USA the death penalty is still used by 36 states. One question concerns whether there is a deterrent effect on murder from having a death penalty.

Ehrlich (1975) used a Becker-type model in which the murderer maximises utility by comparing the costs and benefits of the crime. Time-series data were used for the US for the period 1933–69. Ehrlich represented the benefits of murder by property-crime variables because murder was correlated with property crime. The benefits of murder were determined by data on unemployment, levels of wealth, and the age structure and racial composition of the population. The costs of murder (to the murderer) were measured by alternative measures of the hazard of punishment: the probability of arrest for murder; the probability of conviction given arrest; and the probability of execution given conviction. Ehrlich predicted the relative strength of these deterrents to be in the order given above.

Ehrlich's results showed that the murder rate was negatively and significantly correlated with the arrest, conviction and execution probabilities, with the relative strength of deterrence in this order. One striking conclusion was that one extra execution in a year would deter eight murders, which was famously quoted in the US Supreme Court, in the case of *Gregg v. Georgia* in 1976. A less-quoted result was that the deterrent effect of improved labour-market conditions was greater than for the justice variables.

There are a number of possible criticisms of Ehrlich's (1975, 1977) work. First, does deterrence logically follow from conviction? Capital punishment might make juries less willing to convict, in which case it might not act as a deterrent. There was in fact a fall in the number of murderers found insane by courts in the UK after the abolition of capital punishment in 1965, which suggested the courts were previously looking for excuses not to convict. In this connection, Lempert (1981) re-estimated Ehrlich's model and found that an increase in the use of the death penalty decreased the probability of conviction by 17 per cent.

Secondly, some subsequent researchers found it difficult to replicate Ehrlich's results. One statistical criticism of the work is that its conclusions seem to be sensitive to the mathematical form used. Changing the formulation can change the results, as can changing the time period. Passell and Taylor (1977) re-estimated Ehrlich's model but excluded data for the period 1962–69, in which executions dropped from 47 to zero and crime rose sharply. They found no significant relationship between executions and murder for 1933–61, with the unlikely implication that capital punishment was a deterrent in the USA only for the period after 1962. Leamer (1983) uses a model of capital punishment in his demonstrations of how statistics may be misused to support prior beliefs.

Several studies strongly support Ehrlich's broad conclusions. Wolpin (1978b) replicates Ehrlich (1975) on data for England and Wales for 1929–68 and concludes that one extra execution in a year would have deterred four murders. In more recent work Deadman and Pyle (1993) use the modern technique of intervention (time-series) analysis on data for England and Wales over the period 1880–1989 and for Scotland for 1884–1989. If socioeconomic, demographic and law enforcement variables change slowly over time then murder data should show some inertia, whereas there is a significant shift in the trend for murder after abolition of the death penalty in 1965 equivalent to about 52 extra murders a year for the UK.

Some evidence clearly supports a deterrent effect for capital punishment. It does not follow from this that there is a strong economic argument in favour of using a death penalty. The costs of capital punishment are high, so it is not necessarily cost-effective. In the USA there are high costs attached to running an exhaustive appeal system and operating 'death row'. It may also be argued that the costs of mistakes are very high when an irreversible penalty of this sort is used. Growing concern over the irreversibility of mistaken hangings may well have been decisive in the move to abolition in the UK.

Conclusions

Economic analysis aids our understanding of crime and punishment by providing unique insights. There is statistical evidence that individuals engage in crime as a rational activity and are deterred by changes in criminal justice variables. This appears to be particularly true in the case of property crime. Increases in the probability of eventual punishment appear to have the strongest deterrent effect. Nonetheless, increasing the severity of sentences appears to be the cheapest way to obtain a given reduction in criminal activity. It would appear that if society is serious about tackling modern crime, a toughening of sentencing is likely to be the most successful approach. A final theme is that deterrence may be seen as complementary to other approaches such as tackling poverty or drunkenness.

References

Becker, G. (1968) 'Crime and Punishment: an Economic Approach', *Journal of Political Economy*, 76, pp. 169–217.

Blumstein, A. and Nagin, D. (1977) 'A Stronger Test of the Deterrence Hypothesis', *Stanford Law Review*, 29, pp. 241–76.

Buchanan, C. and Hartley, P. R. (1992) *Criminal Choice: The Economic Theory of Crime and its Implications for Crime Control*, Policy Monograph 24 (St Leonards, New South Wales: Centre for Independent Studies).

Cameron, S. (1988) 'Economics of Crime Deterrence', *Kyklos*, 41, pp. 301–23.

Carr-Hill, R. and Stern, N. H. (1979) *Crime, the Police and Criminal Statistics* (New York: Academic Press).

Cook, P. J. and Zarkin, G. A. (1985) 'Crime and the Business Cycle', *Journal of Legal Studies*, 14, pp. 115–28.

Deadman, D. F. and Pyle, D. J. (1993) 'The Effect of the Abolition of Capital Punishment in Gt Britain: an Application of Intervention Analysis', *Journal of Applied Statistics*, 20, pp. 191–206.

Dnes, A. (1996) *Economics of Law* (London: International Thompson Press).

Ehrlich, I. (1973) 'Participation in Illegitimate Activities: a Theoretical and Empirical Investigation', *Journal of Political Economy*, 81, pp. 521–64.

—, (1975) 'The Deterrent Effect of Capital Punishment: a Question of Life and Death', *American Economic Review*, 65, pp. 397–47.

—, (1977) 'Capital Punishment and Deterrence: some Further Thoughts', *Journal of Political Economy*, 85, pp. 741–88.

Field, S. (1990) 'Trends in Crime and their Interpretation', *Home Office Research Paper 119* (London: HMSO).

Leamer, E. F. (1983) 'Let's Take the Con out of Econometrics', *American Economic Review*, 23, pp. 31–43.

Lempert, R. (1981) 'Desert and Deterrence: an Assessment of the Moral Bases of the Case for Capital Punishment', *Michigan Law Review*, 79, pp. 1177–1231.

Murray, C. A. and Cox, L. A. (1979) *Beyond Probation: Juvenile Corrections and the Chronic Offender* (Beverly Hills: Sage)

Nagin, D. S. (1998) 'Criminal Deterrence Research at the Outset of the Twenty-first Century', *Crime and Justice: A Review of Research*, 23, pp. 1–42.

Passell, P. and Taylor, J. B. (1977) 'The Deterrence Effect of Capital Punishment', *American Economic Review*, 67, pp. 445–51.

Posner, R. A. (1985) 'An Economic Theory of Criminal Law', *Columbia Law Review*, 85, pp. 1193–1231.

Pyle, D. J. (1983) *The Economics of Crime and Law Enforcement* (London: Macmillan).

—, (1989) 'Economics of Crime in Britain', *Economic Affairs*, 9, pp. 6–9.

—, (1993) 'An Economist Looks at Crime in Britain'. Paper given to the European Policy Forum/Social Market Fund.

—, and Deadman, D. F. (1994) 'Crime and Unemployment in Scotland: some Further Results', *Scottish Journal of Political Economy*, 41, pp. 314–24.

Reilly, B. and Witt, R. (1992) 'Crime and Unemployment in Scotland: an Econometric Analysis Using Regional Data', *Scottish Journal of Political Economy*, 39, pp. 13–28.

Shavell, S. (1985) 'Criminal Law and the Optimal Use of Non-monetary Sanctions as a Deterrent', *Columbia Law Review*, 85, pp. 1232–62.

—, (1993) 'The Optimal Structure of Law Enforcement', *Journal of Law and Economics*, 36, pp. 255–87.

Tullock G. (1968) 'The Welfare Costs of Tariffs, Monopoly and Theft', *Western Economic Journal*, 5, pp. 224–32.

Walmsley, J. (1986) 'Personal Violence', *Home Office Research Paper 89*, London: HMSO.

Willis, K. (1983) 'Spatial Variations in Crime in England and Wales', *Regional Studies*, 17, pp. 261–72.

Wilson, J. and Herrnstein, R. (1985) *Crime and Human Nature* (New York: Simon and Schuster).

Withers, G. (1984) 'Crime, Punishment and Deterrence in Australia: an Empirical Investigation', *Economic Record*, 60, pp. 176–85.

Witte, A. (1980) 'Estimating the Economic Model of Crime with Individual Data', *Quarterly Journal of Economics*, 94, pp. 57–84.

Wolpin, K. (1978a) 'An Economic Analysis of Crime and Punishment in England and Wales, 1894–1967', *Journal of Political Economy*, 86, pp. 815–40.

—, (1978b) 'Capital Punishment and Homicide: the English experience', *American Economic Review*, 68, pp. 422–427.

3

Editors' Introduction

Dnes showed the general importance of the deterrence hypothesis and alerted us to some of its specific effects. We saw that it accounts for a considerable range of criminal behaviour but that its impact is greater in property crime, and that it has to be mediated by other considerations if it is usefully to be applied. David Pyle takes us further into the hypothesis with a focus on the impact of economic activity on criminal participation. Pyle addresses crime as a labour-intensive activity, so that criminal participation is seen as a labour supply decision, considerations being earnings in legal employment, the returns from crime and the probability of unemployment. Pyle demonstrates how this perspective can be expressed formally, giving us a testable understanding of the conditions under which individuals are likely to enter criminal activity. There are trade-offs between a number of factors to take into account, but the virtue of a formal model is that their effects can be assessed by changing either one or several factors and seeing how the results produced compare with crime data.

While economic models suggest unemployment and low income would motivate criminal activity, a counterweight is that times of poor economic performance may reduce criminal opportunities. Pyle shows us some of the macroeconomic evidence for a link between crime and unemployment, and for linkage between crime and economic activity as measured by indices such as gross domestic product. The indications are that the relationship between economic activity and crime differs according to time frame: in the short term they move in opposite directions, in the long term they increase together.

Economists, Crime and Punishment
David Pyle

1 Introduction

At first sight crime may appear to be a peculiar topic for economic analysis. What can an economist contribute to an understanding of such a self-evidently non-economic phenomenon? It turns out that economists have a great deal to say on the subject – first, about what motivates individuals to commit crime and second, about how society should allocate resources in order to reduce the damage caused by crime. In this chapter we will focus on the first of these issues.

The interest of economists in the question of crime can be traced as far back as Adam Smith's Lectures on *Justice, Police, Revenue and Arms*, where Smith claimed that 'the establishment of commerce and manufactures is the best police for preventing crimes' (Smith, 1763: 155). However, it was not until the late 1960s that economists began to study crime and punishment at all seriously. This revival of interest owes much to the work of Gary Becker (1968, reprinted in this volume as chapter 1).

Crime is a major social problem. Victim surveys suggest that, in many OECD countries, between 25 and 30 per cent of individuals are the victims of criminal acts every year (*Criminal Statistics England and Wales*, 1997: Table 10.10). Most recorded crime is acquisitive or property related. In England and Wales about 70 per cent of all recorded crime falls into the categories of burglary and theft, whilst another 20 per cent of recorded offences are cases of fraud or criminal damage. Recorded crime has shown a marked upward trend during the post-war period. In England and Wales it increased by a factor of about ten between 1946 and 1996 (see Fig. 3.6). Similar trends have occurred in other countries. As a consequence public expenditure on the criminal justice system has grown substantially; in the UK by more than 50 per cent in real terms in the past 10 years, so that now more is spent on law enforcement than on the whole higher education system. Similar observations could be made about most advanced, industrialised nations.

In this chapter we will explore the economists' model of criminal participation, which is based upon the premise that criminals respond to incentives. We will also examine attempts to test the predictions of that model, particularly in relation to the effectiveness of deterrence variables and the impact of economic activity upon participation in crime.

2 Economics of the Criminal Law

In the economic approach to crime and punishment, criminals are viewed as rational individuals, who allocate their time between legitimate and criminal pursuits (see Section 3). They are assumed to make this choice on the basis of the costs and benefits attaching to these activities. The benefits include the expected rewards of criminal activity, whilst the costs include penalties which might accrue if you are caught and punished. Economists view criminals as responding to incentives, much as consumers respond to prices and workers respond to wages. This may seem odd to criminologists, sociologists and psychologists. For economists it is not the assumption of what motivates criminals that is important, but the predictions which follow from it. If these predictions are tested and found wanting the approach will be rejected and another line of enquiry pursued. The architects of the Criminal Law possibly shared economists' views about criminal motivation, for the penalties for crimes are clearly set in an attempt to alter behaviour (Clarkson, 1995).

In English Law there are two features of a crime. These are (i) Actus reus (the forbidden conduct) and Mens rea (a blameworthy state of mind or intent to harm). Of course, some crimes can be committed without intent. For example, driving a motor vehicle without due care and attention may result in an accident in which someone dies, but the driver did not intend to kill the victim. Nevertheless, the defendant would be charged with a crime on the grounds (presumably) that he ought to have realised that behaving in this way might have disastrous consequences. No doubt the absence of intent is a mitigating factor in reducing the maximum sentence for such a crime below that for murder.

Criminal liability is constructed around the notion of punishing those who are blameworthy in causing harm. To be blameworthy the action must have been voluntary. You cannot be held responsible for involuntary actions caused by being unconscious, asleep or in a state of hypnosis. This has some bearing on the economic approach, in that one justification for not punishing involuntary actions is that they cannot be deterred. Thus the Criminal Law is based on the notion that individuals are responsible for their actions. 'Being a responsible agent means that man is capable of reason... he can thus control his actions and can choose whether to comply with the law or not' (Clarkson, 1995: 52).

Table 3.1 Stealing as a dominant strategy

		B	
		Trade	Steal
A	Trade	10,9	7,11
	Steal	12,6	8,8

On its own blameworthy conduct would appear to be insufficient to merit the involvement of the State in what appears to be a purely private matter between a criminal and his/her victim. What distinguishes a crime from a tort is the public nature of the harm caused. The act does not harm the victim alone, but also adversely affects the well-being of other members of the public. For example, if you are robbed, then I suffer too. (Of course, my suffering may not be pure altruism. The fact that you are attacked may increase my fear that I will be attacked.) As a result, the social costs of crime are larger than its private costs and the State cannot simply leave the prosecution and punishment of crimes to the civil courts. If it were to do so there would be too little punishment and too high a level of crime, because criminals would have to merely compensate their victims for the harm done, which is less than the social costs of the wrong doing (see Dnes, Chapter 2 in this volume).

In the absence of a Criminal Law, stealing may turn out to be a dominant strategy. Suppose that A and B are farmers. A grows cabbages and B raises cattle. They could either trade with or steal from one another. Imagine that we live in a society in which there is no Criminal Law, no police and no courts. We set out in Table 3.1 the hypothetical pay-offs to A and B under these different scenarios.

The numbers in the cells are profits under the different strategies. The numbers are not entirely arbitrary. They have been selected to illustrate a particular point. In practice they may be different from those shown. The first number in each cell is A's profit, the second is B's. If both refrain from stealing, joint profits are 19, which are shared 10 to A and 9 to B. However, if B devotes her time to stealing and not raising cattle, she can make 11. If A refrains from stealing she now only makes 7. A's loss exceeds B's gain, because A has to put resources into protecting her cabbages. Likewise if A steals but B does not, A gains 2 and B loses 3. In this game stealing is a dominant strategy. For A, 12 > 10 and 8 > 7, so no matter what B does A should steal. Likewise for B, 11 > 9 and 8 > 6, so that B should steal, whatever A does. A and B will end up with joint profits of 16, rather than the 19 they would get if they traded. They are clearly worse off, but are driven to this position in the pursuit of their selfish interests. This example is a case of the so-called Prisoner's Dilemma, which permeates a great deal of economists' thinking. This is a situation in which the individuals are driven to choose a set of strategies which are collectively inferior. However, whilst they may perceive the problem they cannot commit each other to the preferred outcome.

Of course, A could threaten to retaliate against B if B were to steal. B would then have to compare her short-term gain (2 for one period only) against a long-term loss (1 for ever more). This threat may work in a two-person 'game', but it would be extremely difficult for A (and B) to reach agreements of this kind with all potential thieves in a multi-person society. The transaction costs of reaching a large number of bilateral agreements would be prohibitive. So, it may be more efficient to have a legal system which imposes penalties upon those who transgress its code, the aim being to discourage stealing by raising the costs and reducing the benefits of engaging in such activity.

Why cannot tort law deter such activities? One explanation is that the damages required might exceed the offender's ability to pay. We only have to consider the case of murder to see why this might be. There may be another explanation. Suppose that you own a painting which you value at £10 000, but I value it at £15 000. Rather than pay you for it, I decide to steal it. If the probability of your obtaining a judgement against me is 1, then granting you damages of £10 000 is insufficient to dissuade me from stealing your painting. Even granting damages of £15 000 is only sufficient, if I am risk neutral, to make me indifferent between stealing your painting and buying it from you. If the probability of being caught and punished is less than 1, then penalties for theft would need to be punitive. For example, if I believe that the probability of my being caught is only one in ten, then the penalty for stealing the painting needs to be at least £150 000 in order to deter me (Posner, 1985, and Dnes, Chapter 2 in this volume). Of course, if my wealth is less than £150 000 I may have to go to prison.

The Criminal Law assumes that 'man' is a moral agent who is responsible for his actions. Economists further assume that these actions can be altered by appropriate structuring of incentives, but economic analysis suggests that criminal acts cannot be optimally deterred by compensatory damages. This is for two reasons – first, crime involves an external cost which raises social costs above private costs, and second, compensatory damages are in any case too small to deter offenders.

In the next section we set out a more formal theoretical model of criminal behaviour which rests on the assumption that potential criminals are capable of making rational choices about whether or not to engage in such acts.

3 The Economic Approach to Criminal Participation

3.1 Theory

The economic approach to criminal involvement rests on the assumption that most potential criminals are rational people who respond to incentives. Becker (1968) argued that individuals will commit a crime if the benefits (expected utility) from committing the crime exceed those derived from legitimate activity. In Becker's model no account is taken of the disutility (qualms of conscience) which individuals might incur by engaging in criminal activity. (For an attempt to incorporate such moral scruples see Block and Heineke, 1975.) If you find the economists' approach hard to swallow, then recall that the vast majority of crime is property–related and possibly financially motivated.

An important aspect of criminal activity is that it is inherently risky. After all, you might get caught, which would reduce the expected utility from criminal acts. As a result, Becker claimed that potential criminals will be deterred from committing crimes by increases in (i) the probability of being caught and punished and (ii) the amount of punishment if caught, because each of these reduces expected utility.

A Simple Model of Criminal Activity

In Becker's model individuals are treated as if they are gambling with a part of their wealth, but it might be more helpful to consider crime as a labour-intensive activity. For example, planning and carrying out a burglary or robbery may take time. In effect this means that criminal participation is treated as a labour supply decision and is influenced by factors such as earnings in legitimate work, returns to criminal activity and the probability of unemployment.

Below we set out a simple model in which individuals choose how to allocate their time (T) between legitimate activity (t_l) and crime (t_c). The wage rates in the two forms of activity are fixed and given by W_l and W_c, respectively. If caught, offenders are punished by a penalty (f) which is related to the amount of time they spend in criminal activity. What income level the individual enjoys will depend upon whether or not s/he is caught and convicted. Call these income levels I_u (for unsuccessful) and I_s (for successful). Hence,

$$I_s = W_l t_l + W_c t_c$$
$$I_u = W_l t_l + W_c t_c - f t_c = I_s - f t_c$$

If someone devotes all of her time to legitimate activity, then her income will be the same whether she is caught or not, so that

$$I_u = I_s = W_1 T$$

On the other hand if she spends all of her time engaged in criminal activity, then

$$I_u = (W_c - f)T$$
$$I_s = W_c T$$

We draw the opportunity locus in Fig. 3.1. At point A the individual specialises in legitimate activity, i.e. $t_1 = T$, whereas at point B the individual specialises in criminal activity and $t_c = T$. Points between A and B represent non-specialisation in either legitimate or criminal activity. For example, at C (the mid-point of the line AB) the individual devotes half her time to each form of income-generating activity. The closer someone is to A, then the less time that person devotes to criminal activity. The closer such people are to B the more time they spend in criminal activity. The opportunity locus is linear because we have assumed that W_c, W_1 and f are all constant.

Expected utility is given by

$$EU = p\, U(I_u) + (1 - p)U(I_s)$$

Note that p is assumed to be independent of t_c. The preferences define a set of indifference curves and along each indifference curve expected utility is held constant. As both I_u and I_s are desirable, the indifference curves will be negatively sloped and convex to the origin. (In fact, convexity requires the assumption of risk aversion; see Pyle, 1983: 19.)

According to this model, someone will enter criminal activity when the slope of the opportunity locus is flatter than the slope of the indifference curve passing through A in Fig. 3.1. This will arise when $W_c - W_1 > pf$. That is if the increased financial return ($W_c - W_1$) from a unit of time spent in criminal activity outweighs the expected punishment (pf). For those individuals for whom this applies an optimum allocation of time between criminal and legitimate pursuits will be established where the highest possible indifference curve is tangential to the opportunity locus – see point C in Fig. 3.2. This optimum allocation (t_c^*, t_1^*) will depend upon W_1, W_c, p and f as well as their preferences. The unemployment rate can be added to this list if it is treated as a risk of engaging in legitimate activity. We do not develop this point formally, in order to keep the analysis simple. An intuitive explanation is the following. Unemployment lowers

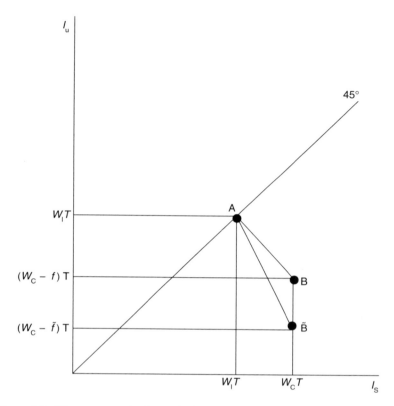

Figure 3.1 The opportunity locus for legitimate and criminal activity

the return from legitimate activity and so an increase in the unemployment rate, *ceteris paribus*, makes criminal activity more attractive. As a result a rational criminal will devote a larger proportion of his time to crime if the unemployment rate rises.

Briefly we will look at the effect of changes in f and p. An increase in the severity of punishment (f) does not affect the co-ordinates of point A in Fig. 3.1, but does change the co-ordinates of point B. In particular, it causes B to drift downwards vertically to ($\tilde{B}$) (see Fig. 3.1). The increase in f clearly incorporates two effects – a substitution effect (the slope of the opportunity locus changes), and an income effect (the locus shifts inwards towards the origin). Both the substitution and income effects cause the individual to spend less time in criminal activity. Of these the income effect perhaps needs some explanation. In the literature on risk

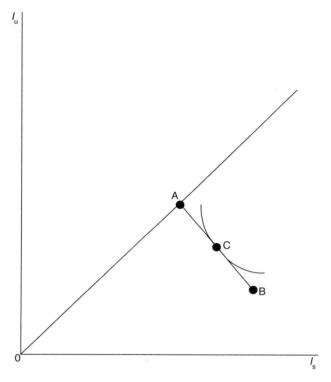

Figure 3.2   Optimum allocation of time between legitimate and criminal activity

taking it is generally assumed that individuals are more willing to accept a risk of any given size as their income increases. In this case the increase in punishment makes them worse off, so they respond by reducing the amount of risk they are prepared to take, i.e. they do less crime.

An increase in the certainty of punishment (p) has no effect on the opportunity locus, but changes the slopes of the indifference curves. If p increases, this has the effect of flattening the indifference curve. That is, for given I_s and I_u the indifference curve passing through the previously optimal point will be less steep, so that the previous optimum can no longer hold. In fact the optimum moves closer to A, i.e. involves less time spent in criminal activity; see Fig. 3.3.

This was not understood by Dau-Schmidt (1990), who argued that 'economists analyse only the opportunity-shaping method of affecting

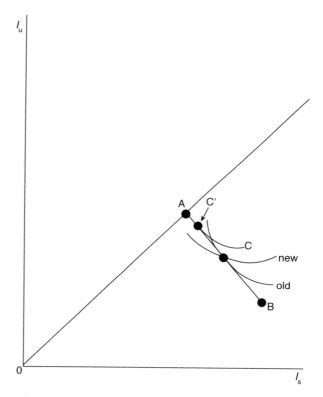

Figure 3.3 Changes in the certainty of punishment and the indifference curve

individual behaviour, assuming that individual preferences are exogenous and immutable' (p. 5).

3.2 Evidence on Deterrent Effects

From the standpoint of criminal justice policy, it is important to know the approximate quantitative effects which the criminal justice variables (certainty and severity of punishment) have upon crime rates. As a result, a great deal of research effort has been devoted to the estimation of so-called crime 'supply' equations.

Most of the empirical testing of the relationship between crime and deterrence variables has analysed data on recorded crimes using standard statistical procedures, normally multiple regression analysis. Three basic types of study can be distinguished. These are (i) cross-section comparisons of areas at a point in time, such as either provinces or counties in a particular year; (ii) analyses of a particular area (region, country) over a period of time; and (iii) longitudinal studies of individuals, usually released prisoners.

The use of recorded crime statistics presents some difficulties. It is well known that these statistics are error-prone and so how reliable are results of statistical modelling which rely upon using recorded crime statistics? The problems may be further compounded if one of the deterrence variables is either the conviction rate or arrest rate which is obtained by dividing the number of convictions or arrests by the number of recorded crimes. This will bias the conviction/arrest rate upwards and impart a negative bias to the observed deterrent effect of the certainty of punishment (see either Pyle, 1983 or Lewis, 1986 for a full discussion of this). In practice, measurement error may not be as big a problem as it appears. For example, Pudney, Deadman and Pyle (1999) have used data, from the General Household Survey and the British Crime Survey, on the reporting of burglary offences to examine the impact of measurement error in a generic model of crime. Simulation experiments reveal that 'the general nature of measurement error biases in the model of crime is not particularly serious' (p. 23). So, studies which have ignored measurement error problems may still produce meaningful conclusions.

The empirical literature attempting to assess the deterrent effect of punishment is now vast and has been surveyed in a number of places (see, for example, Pyle, 1983; Lewis, 1986; Cameron, 1988). Many of these studies lend support to the view that crimes are deterred by increases in either the likelihood or the severity of punishment. However, there is rather less consensus about the size of any deterrent effect. This variation may be due to different data sets. It may also be because of differences in model specification. Also, it is clear that some studies of deterrent effects have omitted variables to measure the impact of positive incentives such as employment and earnings opportunities (Ehrlich, 1996).

Most of this work has used data for the USA and only a few studies have used data for other countries, e.g. England and Wales, Canada and Finland. Whilst British studies (Wolpin, 1978; Carr-Hill and Stern, 1979; Willis, 1983; Pyle, 1989; Reilly and Witt, 1996) have all used slightly different sets of statistics and rather different model specifications, they all support the predictions of the economic model concerning deterrent effects. These studies also indicate that the deterrent effect of certainty of punishment is usually stronger than the deterrent effect of its severity. Where the crime index has been disaggregated, it seems that punishment may exert a stronger deterrent effect for property crimes than it does for violent or sexual offences (Willis, 1983). Also, within the general group of crimes against property – that is burglary, theft and robbery – it has been shown that deterrent effects differ quite significantly from one crime group to another (Pyle, 1989).

3.3 Crime and the Economy

On the whole, unemployed people tend to be less happy (or more stressed) than their employed counterparts (Clark and Oswald, 1994). Empirical research has attempted to link various social ills (suicide, illness and crime) to unemployment (Hakim, 1982). Nevertheless the idea that crime may be related to unemployment has proved particularly controversial. Economic models of criminal behaviour clearly suggest that factors such as unemployment and low income might motivate some individuals to engage in crime. On the other hand, in times of relative economic decline there may be fewer opportunities for potential criminals (for a discussion of opportunity versus motivation see Pyle and Deadman, 1994).

In the 1980s several extensive literature reviews of the relationship between crime and unemployment (Freeman, 1980; Tarling, 1982; Long and Witte, 1983; Box, 1987) reached the conclusion that the relationship is at best inconsistent. Chiricos (1987: 188) has referred to this as 'the consensus of doubt', whilst Cantor and Land (1985) argue that this finding can be explained by the countervailing effects of criminal motivation and criminal opportunity. Nevertheless, most econometric studies of the relationship between crime and unemployment in Britain have found that unemployment has a significant, positive, although small, effect upon recorded crime (see Wolpin, 1978; Willis, 1983; Pyle, 1989; Hale and Sabbagh, 1991; Reilly and Witt, 1992, 1996; Witt, Clarke and Fielding, 1998). A notable exception to this rule is the study by Carr-Hill and Stern (1979). Interestingly, none of these articles has attempted to test whether there is a link between crime rates and either youth unemployment or long-term unemployment. In general they have focused on the overall unemployment rate.

Recently attention has switched from studying the relationship between crime and unemployment to consider instead the role of other economic indicators, particularly consumers' expenditure (Field, 1990) and gross domestic product (Pyle and Deadman, 1994). Recorded unemployment may no longer be a good economic indicator for the UK, especially following the substantial revisions and redefinitions which have occurred throughout the 1980s. Also, it is well known that

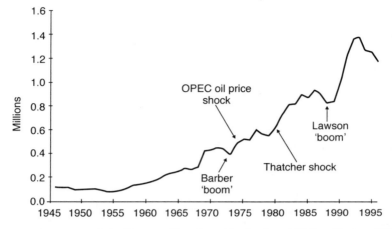

Figure 3.4   Recorded offences of burglary, England and Wales, 1946 to 1996

unemployment lags behind the cycle in economic activity, by on average between six and 12 months but sometimes by as much as two years. As a consequence, when the economy turns into a recession, unemployment will not necessarily be rising. However, if property crime is a response to worsening economic circumstances, then it will already have begun to increase as reduced income and short-time working begin to reduce living standards (Allan and Steffensmeier, 1989).

The possible link between recorded crime and the state of the economy can be seen perhaps more clearly in Fig. 3.4, which charts numbers of recorded offences of burglary in England and Wales from 1946 to 1996. There is clearly a strong upward trend in these data. However, there have been years in which recorded crime has fallen, e.g. 1973, the late 1970s, the late 1980s and every year since 1993. Some of these fluctuations in the series appear to coincide with particular economic events. For example, the fall in recorded crime in 1973 followed the 'Barber boom' of 1972–73 when real GDP grew by more than 7 per cent in a single year and the rise in recorded crime in 1974–76 followed the OPEC oil price shock of 1973–74.

It would seem that the relationship between recorded crime and economic activity is different in the long term, when both increase together, and in the short term, when they move in opposite directions. This is shown schematically in Fig. 3.5.

The short-term link between recorded crime and economic activity may be more apparent if we focus on changes in recorded crime and changes in

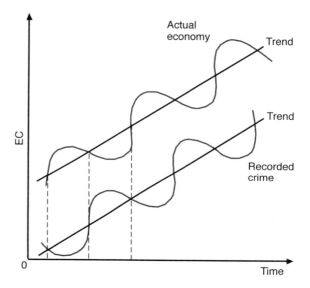

Figure 3.5 Long-run/short-run crime–economy relationships

real GDP. In Fig. 3.6 we have plotted annual percentage changes in recorded offences of burglary and real GDP between 1986–87 and 1995–96. There would seem to be a clear negative association between these phenomena. As real GDP increases the rate of increase in recorded

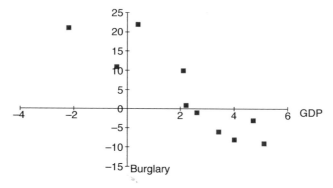

Figure 3.6 Annual percentage changes in real GDP and recorded burglary, 1986/87 to 1995/96

burglary slows and if real GDP increases fast enough recorded offences of burglary actually fall. Of course, there may be other explanations for these changes and the apparent link between economic change and movements in recorded crime could be coincidental. Recent research has tried to address this issue.

These relationships have been explored in some detail by Field (1990), Pyle and Deadman (1994), Osborne (1995), Deadman and Pyle (1997), Hale (1998) and Pudney *et al.* (1999). From these recent studies it would appear that recorded crime, and particularly recorded property crime, is much more closely correlated with the level of economic activity in England and Wales, at least in the short term, than had previously been thought. These studies, incorporating developments in the econometric analysis of time-series data, have reaffirmed a much earlier conclusion reached by Thomas (1927) and Henry and Short (1954).

4   Conclusion

In the 30 years since Gary Becker introduced the notion of an economic approach to crime and punishment, economists have devoted a considerable amount of time and energy to studying what appears at first sight to be an issue that is not susceptible to economic analysis. The economic approach rests upon the assumption that criminals are rational individuals who respond to incentives. Much of the literature on the economics of crime and punishment has been devoted to testing empirically the implications of that assumption. The assumption itself may not be so unreasonable, for there is clear evidence that the Criminal Law attempts to establish incentives for individuals to behave in particular ways. However, that itself is not a test of the adequacy of the economic approach.

Economists have attempted to test that model directly by estimating relationships between crime and various criminal justice variables and labour market indicators. Whilst this volume of work has been impressive, it has yet to furnish definitive conclusions. Whilst there is a great deal of evidence that crime is inversely related to measures of punishment, there is less consensus about the size of any deterrent effect. The evidence on the relationship between crime and economic factors is even less clear-cut, although recent work with time-series data in the UK has begun to furnish more convincing evidence than has been produced before. The difficulty of arriving at a consensus on the effects of economic factors upon crime arises in part because the different types of study (time-series or cross-section) pick up different aspects of the labour market, i.e.

demand-side or supply-side influences. A further problem in interpreting the results of empirical work is caused by the fact that some studies examining the impact of sanctions variables on crime have ignored the effects of economic factors, whilst some studies of the effect of economic forces have ignored criminal justice variables (Ehrlich, 1996). As the approach suggests that crime is determined by both positive and negative incentives, this is an unfortunate omission. On the other hand, the problem of measurement error may not be quite as severe as was once thought.

Clearly a great deal of progress has been made by economists in understanding the causes of crime, although more research could be done. However, almost as important a task is to convince scholars from other disciplines of the usefulness of the economic approach.

References

Allan, E. A. and Steffensmeier, D. J. (1989) 'Youth, Underemployment and Property Crime: Differential Effects of Job Availability and Job Quality on Juvenile and Adult Arrest Rates', *American Sociological Review*, 54, pp. 107–23.

Becker, G. S. (1968) 'Crime and Punishment: an Economic Approach', *Journal of Political Economy*, 76, pp. 169–217.

Block, M. K. and Heineke, J. M. (1975) 'A Labor Theoretic Analysis of Criminal Choice', *American Economic Review*, 65, pp. 314–25.

Box, S. (1987) *Recession, Crime and Punishment* (London: Macmillan).

Cameron, S. (1988) 'The Economics of Crime and Deterrence: a Survey of Theory and Evidence', *Kyklos*, 41, pp. 301–23.

Cantor, D. and Land, K. C. (1985) 'Unemployment and Crime Rates in Post-World War II United States: A Theoretical and Empirical Analysis', *American Sociological Review*, 50, pp. 317–32.

Carr-Hill, R. A. and Stern, N. H. (1979) *Crime, the Police and Criminal Statistics* (New York: Academic Press).

Chiricos, T. G. (1987) 'Rates of Crime and Unemployment: an Analysis of Aggregate Research Evidence', *Social Problems*, 34, pp. 187–212.

Clark, A. E. and Oswald, A. J. (1994) 'Unhappiness and Unemployment', *Economic Journal*, 104, pp. 648–59.

Clarkson, C. M. V. (1995) *Understanding Criminal Law* (London: Sweet & Maxwell).

Dau-Schmidt, K. (1990) 'An Economic Theory of Criminal Law', *Duke Law Journal*, pp. 1–38.

Deadman, D. F. and Pyle, D. J. (1997) 'Forecasting Recorded Property Crime using a Time-series Econometric Model', *British Journal of Criminology*, 37, pp. 437–45.

Ehrlich, I. (1973) 'Participation in Illegitimate Activities: a Theoretical and Empirical analysis', *Journal of Political Economy*, 81, pp. 521–64.

—, (1996) 'Crime, Punishment and the Market for Offences', *Journal of Economic Perspectives*, 10, pp. 43–67.

Field, S. (1990) *Trends in Crime and their Interpretation: A Study of Recorded Crime in Post-war England and Wales*. Home Office Research Study 119.

Freeman, R. B. (1980) 'Crime and Unemployment', in J. Q. Wilson (ed.), *Crime and Public Policy* (San Francisco: ICS Press) pp. 89–106.

Hakim, C. (1982) 'The Social Consequences of High Unemployment', *Journal of Social Policy*, 11, pp. 433–67.

Hale, C. (1998) 'Crime and the Business Cycle in Post-war Britain Revisited', *British Journal of Criminology*, 38, pp. 681–98.

—, and Sabbagh, D. (1991) 'Testing the Relationship between Unemployment and Crime: a Methodological Comment and Empirical Analysis Using Time-series Data for England and Wales', *Journal of Research in Crime and Delinquency*, 28, pp. 400–17.

Henry, A. F. and Short, J. F. (1954) *Homicide and Suicide* (Glencoe, IL: Free Press).

Lewis, D. E. (1986) 'The General Deterrent Effect of Longer Sentences', *British Journal of Criminology*, 26, pp. 47–62.

Long, S. K. and Witte, A. D. (1983) 'Current Economic Trends: Implications for Crime and Criminal Justice', in K. N. Wright (ed.), *Crime and Criminal Justice in a Declining Economy* (Cambridge, MA: Oelgeschlager, Gunn & Hain) pp. 69–143.

Osborne, D. (1995) 'Crime and the UK Economy', Robert Schuman Centre Working Paper 95/15, European University Institute.

Posner, R. (1985) 'An Economic Theory of the Criminal Law', *Columbia Law Review*, 85, pp. 1193–1231.

Pudney, S., Deadman, D. F. and Pyle, D. J. (1999) 'The Relationship between Crime, Punishment and Economic Conditions: Is Inference Reliable when Crimes are Under Recorded?', *Journal of the Royal Statistical Society*, Series A (Statistics in Society).

Pyle, D. J. (1983) *The Economics of Crime and Law Enforcement* (London: Macmillan).

—, (1989) 'The Economics of Crime in Britain', *Economic Affairs*, 9, pp. 6–9.

—, and Deadman, D. F. (1994) 'Property Crime and the Business Cycle in Post-war Britain', *British Journal of Criminology*, 34, pp. 339–57.

Reilly, B. and Witt, R. (1992) 'Crime and Unemployment in Scotland', *Scottish Journal of Political Economy*, 39, pp. 213–28.

—, (1996) 'Crime, Deterrence and Unemployment in England and Wales: An Empirical Analysis', *Bulletin of Economic Research*, 48, pp. 137–55.

Smith, A. (1763) *Lectures on Justice, Police, Revenue and Arms* (1763).

Tarling, R. (1982) 'Crime and Unemployment', *Home Office Research and Planning Bulletin*, no. 12.

Thomas, D. S. (1927) *Social Aspects of the Business Cycle* (New York: Gordon & Breach).

Willis, K. G. (1983) 'Spatial Variations in Crime in England and Wales: Testing an Economic Model', *Regional Studies*, 17, pp. 261–72.

Witt, R., Clarke, A. and Fielding, N. (1998) 'Crime, Earnings Inequality and Unemployment in England and Wales', *Applied Economics Letters*, 5, pp. 265–7.

Wolpin, K. I. (1978) 'An Economic Analysis of Crime and Punishment in England and Wales, 1894–1967', *Journal of Political Economy*, 86, pp. 815–40.

4

Editors' Introduction

With the seminal contribution of Becker, and Dnes' and Pyle's elaborations based on it, we now have a good idea of the tools economists apply in explaining crime and assessing punishment. Economic ideas can also inform more discursively oriented efforts to understand the criminal world, efforts directed as much to understanding an alien culture as to practical application in, for example, law enforcement.

Such is the case in the influential work of Diego Gambetta and Peter Reuter on the mafia, work which draws variously on ethnography, the economic theory of the firm, social history, criminology, and the theory of state regulation. Sometimes, as a result of a new insight, the hitherto-unexplained suddenly assumes sense, so that afterwards one wonders how one can ever have thought things to be otherwise. From Gambetta and Reuter we come to understand that the primary beneficiaries of the mafia (and other syndicate-maintaining organisations) are the owners of the businesses being 'coerced'. Social scientists have developed the concept of 'latent functions' to account for the unanticipated effects of social processes and social institutions; Gambetta and Reuter show how economics can be a useful means to identify such functions (note that a precursor of the argument Gambetta and Reuter develop can be found in Becker's discussion of collusive business practice).

Gambetta and Reuter argue that the key protection racketeers supply to legitimate businesses is the enforcement of trade allocation agreements among independent firms, and rather than monopolistic restraint of trade, the interest of the mafia as cartel enforcer lies in maintaining a reasonable diversity of participating businesses. In terms of understanding the coexistence of the business world and the world of organised crime, Gambetta and Reuter's use of some fundamental theoretical tools of economics offers real analytic purchase.

Conspiracy among the Many: the Mafia in Legitimate Industries*

Diego Gambetta and Peter Reuter

1 Introduction

This chapter considers the modes by which the mafia exercises its influence on a number of legitimate industries in both Sicily and the United States. In particular it discusses the kinds of service the mafia provides, the economic consequences of its influence, the conditions that induce the entry of the mafia in specific industries, and the conditions and policies that make it disappear. We share the view that mafia protection in legitimate industries, although occasionally rapacious and unreliable, is frequently neither bogus nor limited to intimidating new entrants. Under some (perhaps most) circumstances, the primary beneficiaries are the owners of the firms being coerced.

This view is based both on theoretical arguments and empirical evidence. The chapter relies on a series of case-studies which were autonomously developed by the authors and presented elsewhere (Reuter, 1987, 1993; Gambetta, 1993). The most important type of protection supplied by racketeers to legitimate industries, which is found in most industries we studied, is the enforcement of a variety of allocation agreements among independently owned firms, with racketeer income as payment for the service. This runs counter to two widespread conceptions of the involvement of the mafia in legitimate industries. One sees the racketeers' role as mere intimidation: the mafioso approach to 'regulated' competition would amount to thugs resorting to muscular persuasion at the expense of an innocent competitor on behalf of either a single monopolist or the mafia's own enterprise. The other sees racketeers as extortionists imposing their presence upon harmless dealers who would rather do without. Let us consider them in turn.

*This chapter was originally published in Fiorentini and Peltzman (eds), *The Economics of Organized Crime* (1995) and is reproduced by kind permission of Cambridge University Press.

When the mafia looks after the interests of a monopolist, it is clearly useless as a cartel enforcer and protection is limited to the intimidation of new entrants. This, however, is seldom the case: although protectors have an interest in not expanding the number of protected firms indefinitely, they are subject to an opposite drive to increase it. First, the larger the number of firms the higher is the mafia's autonomy from and bargaining power over any single one of them (Gambetta, 1993: 22–3, 85). Second, supporting a small number of firms is risky because it provides disappointed competitors with an incentive to seek protection from police or rival racketeers (Reuter, 1987: 6). Third, larger collusive agreements hold competition in check more effectively both because the cartel acts as a discouraging signal of determination and stability, thus reducing the need to threaten potential competitors, and because the option to exit the cartel becomes less attractive for the participants (Fiorentini, 1994).

Although in particular circumstances collusion is self-enforcing (Friedman, 1983: 65, 132 ff), whenever cartels rely on agreements, defection is a potential hazard and a genuine demand for protection may develop. Each partner must feel confident that all other partners will comply with the pact; otherwise the cartel collapses and competition creeps back in. Furthermore, even though anti-competitive outcomes can emerge without cooperation among firms – a large firm may independently decide to exclude new entrants even if rival firms share the advantages but not the cost – whenever restrictive practices require the contribution of all firms, an enforcing agency may be necessary to deter members from free-riding. There is therefore no theoretical reason to expect that the role of the mafia will be one of extortion rather than authentic protection. (Whether the price of protection is 'extortionate' is a separate question to which we return later in Section 8.)

Our argument is that the mafia solves a problem of potential cartels. It may be invited in by entrepreneurs themselves looking to organise some agreement or may initiate the activity itself; we have examples of both. Its comparative advantage is likely to be in organising cartel agreements for large-number industries, as well as making cartels more stable; success, however, requires a number of other conditions, that we spell out in the course of the chapter. Moreover, the mafia has a unique asset in this capacity, namely its reputation for effective execution of threats of violence; this creates a reputational barrier to entry.

We do not, however, claim that large-number cartel organising is the exclusive role of the mafia in legitimate industries. The garment trucking and waterfront industries in the United States (Reuter, 1987: Ch. 3) provide instances of other kinds of predation. However, cartel organisation is an important and distinctive activity of the mafia,[1] and not just in legitimate industries. Although illicit markets do not concern us here, there are signs that the mafia can act as cartel organiser, for instance, in bookmaking in the United States and in tobacco smuggling and purse snatching in southern Italy (Reuter, 1983: 42–4; Gambetta, 1993: 227–35).

2 The Form of Cartel Agreements

The central problem for cartels is to design agreements which can be maintained over time. Three aspects of agreements, which are relevant to human cooperation generally, are also relevant for establishing cartels: the rules to be followed by members, the means of detecting rule violators, and the sanctions to be imposed against violators. Weakness in any of the three is likely to lead to defections and breakdown of the cartel (Stigler, 1964).

2.1 The Rule

At least three dimensions of the rule itself affect cartel members' willingness to enter and remain in association. First, the rule must produce results that, over time, are perceived as equitable, so as to lower risk of defection and complaint. Second, the agreement must be compatible with continued entrepreneurial autonomy. Loss of autonomy – as might arise if the cartel decided that production facilities must be centralised to maximise joint profits – transforms the bargaining positions of members and those who lose production facilities become dependent on others' willingness to honour the original commitment. This may increase the demand on trust to breaking point. Third, because of illegality, there is likely to be a strong inclination to avoid rules which require 'side-payments' among members that could provide valuable evidence for a prosecutor. More generally, illegal agreements are likely to minimise the number of transactions necessary for their effective execution.

2.2 Detection

The second important issue is the speed and cost of detecting violation. The longer it takes to detect it, the greater the return to violation and the lower the expected lifetime of the cartel (Stigler, 1964). Similarly, the profitability of the cartel will be lowered by high costs of monitoring. More generally, the higher the cost of policing agreements the lower the likelihood that cooperation will emerge.

2.3 Sanctions

The final consideration is the nature of available sanctions. Members will doubt a cartel's potential stability if violations of its rules cannot be punished without ending the cartel. A classic difficulty of price-fixing agreements is that efforts to punish price cutters lead to price reductions for all customers. The possibility of selective and cheap sanctions will increase stability.[2]

A limited number of rules can sustain collusive agreements. These are price fixing, output quotas and market sharing. Market sharing seems generally easier to enforce than, and logically prior to, the other two alternatives. In the New York carting industry,[3] for instance, in which individual customers are allocated on the basis of who served them first, cartel members have complete price autonomy. Being fixed in location, customers have little choice if other carters accept the basic rule. By contrast, if price fixing is practised – assuming that the product is homogeneous – firms usually need further allocative measures. Price fixing creates an incentive to resort to covert inducements, such as advertising, gifts, special offers, guarantees, sponsorship, etc. in order to attain higher market shares. Thus, unless firms also have market-sharing rules, price fixing is likely to be unstable, even ignoring potential entry.

3 A Comparison of the Sicilian and New York Mafias

The differences between the mafia in Sicily and New York are several. For example, in Sicily there is a clear geographic division; individual families have specific areas in which they have sole operating authority. Members of other families can operate there only with permission. Within New York the nature of the division is harder to determine. It is certainly not territorial, perhaps reflecting the lack of effective small-area local government that would permit the use of corrupt government authority to establish monopolistic criminal enterprises. Above all, in a highly diversified and complex urban situation such as that of New York, a monopoly of protection on all transactions is hardly conceivable, even in a relatively small area. By contrast, in a village or district of Sicily, with few inhabitants, any one mafia family can still take care of its clients in most of their businesses; but as the number of exchanges multiplies it becomes increasingly laborious to supply territorially based protection, and protection itself tends to become functionally specialised. Historic roles certainly have an influence in determining where (functionally) a family operates in New York. The Lucchese and Gambino families are the only ones of the five New York families that have a presence in the carting industry, just as it is the Genovese that are most active on the docks in Manhattan. That seems to have been the case for over 40 years in the case of carting and nearly 70 years for the docks.

The capacity of the mafia to play a role in cartel organisation is probably affected by this general organisational characteristic. Sicilian industry agreements may have territorial limits because the families cannot operate outside those areas. Indeed, it appears that some sharing agreements are cumbersome to execute because of the territorial division of the mafia families; the queueing for contract bids, discussed below, is an instance where arrangements have to be worked out among different families. Also, in New York there is much more of an explicit market in property rights. Carters buy and sell customers at prices that seem to represent the capitalised value of the customer allocation agreement; it is unclear whether mafiosi levy a tax on these sales. In Sicily there is no evidence of markets in such rights. Another difference is one of scale of individual families. The mafia of Sicily are approximately 3000, in a total of nearly 100 families. In contrast there are an estimated 5000 mafia in New York, but grouped into only five families. Entry is more restricted in New York and the rewards of membership correspondingly greater. Few seem to enter American mafia families before their late thirties, while in Sicily members enter in their twenties.

A further important difference for our purposes is the role of the labour union. In the United States it has been the central instrument by which the mafia has acquired power in particular industries. The union provides an ideal means to dress up extortion or the suppression of competition as the expression of concern for the rights of individual workers. Interestingly, the union is critical for the initiation of a customer allocation agreement but appears not to be so central for its continuation; expectation and reputation may suffice at that point. In Sicily, on the other hand, the union is not a major institution. The mafia has much more direct ties to the political system and, as argued at length in Gambetta (1993), serves to provide quasi-governmental services that the elected government fails to deliver.

One final difference is worth mentioning. Illegal markets have traditionally been a more important source of income to the New York mafia than to the Sicilian mafia. In markets for horse-betting, certain kinds of lottery, loan-sharking and (in earlier eras) prostitution and bootlegging, the mafia has served both as a provider of the goods themselves and as guarantors of contracts and dispute settlers. In recent decades it is the latter role that has been most significant (Reuter, 1983). There are thus parallels between the mafia's current roles in legal and illegal markets. For the Sicilian mafia, although the role of protection has been salient for a long time, markets in illegal commodities have until recently played a less important role. The growth of the heroin market since 1970 has changed that, but the Sicilian mafia in this case has played a direct role in the provision of credit and part of the service too (Gambetta, 1993: 234–44).

4 Instances of Mafia Cartels

With only a modest number of case-studies available, even including those that have appeared in historical studies, we make modest claims as to the generality of the following summary of the prevalence of different kinds of mafia cartel rules.

We know of few settings in which the mafia has enforced *price-fixing* agreements over long periods of time. In the wholesale fruit and vegetable market in Palermo, for example, the dramatic increase in numbers of authorised market middlemen, known as *commissionari*, made it impossible to continue to fix prices. Until 1955 only 12 *commissionari* were operating and price fixing and quota control were feasible because easy to police. A 'man of respect', himself a *commissionario*, acted as guarantor of these arrangements and controlled the largest share of the transactions. Now that the *commissionari* are 77,[4] they say that it would no longer be profitable to pay someone who, like 'Don Peppe' in bygone days, enforced collusion. There are simply too many *commissionari* and, given the large number of customers that use the market every day, someone somewhere would always breach the deal. In Section 7 we provide an analysis of why the numbers grew so substantially and broke the cartel.

Output *quotas* are another important alternative. The little available evidence suggests that quota agreements require extensive and intrusive inspection for their maintenance, and these are not attractive for an illegal market. The only known case of regulating output quotas involving the mafia dates back to the nineteenth century. Franchetti, an early and insightful scholar of the phenomenon, reports the existence of two 'societies' based near Palermo: one of millers – la società Mulini – which we would call a cartel, and the other – la società della Posa – a union of cart-drivers and apprentice-millers. Both these societies are said by Franchetti to have been under the protection of 'powerful mafiosi'. Members paid a fee to the society and agreed not to compete. They kept the price of flour high by regulating output, taking turns to restrict production and receiving appropriate compensation. (Note that current EU agricultural policies are not too dissimilar.) The *capo*-mafia ensured that everybody paid their dues, that the miller whose turn it was to under-produce did not free-ride on his fellows by producing more than he was supposed to; and that the others did not free-ride on him, by failing either to pay the agreed compensation or to restrict production when their turn came (Franchetti, [1876] 1974: 6–7 and 96).

The remaining collusive rule, namely *market-sharing*, can follow any one of three cases: locations, customers or queues. Which of them is adopted depends on the general conditions of the market – physical and technological constraints, size of units of demand and supply, frequency of transactions – and on which proves more suitable to monitor defectors. In what follows we assume for simplicity of exposition that agents sharing the market are sellers, but the same arguments apply to buyers. For instance, in the wholesale fish market in Palermo, middlemen share suppliers, i.e. fishermen.

4.1   Location or Territory

A gets the north side, B the south side; A runs on route X and B on route Y. Trading within territorial boundaries is found in a variety of industries, such as bus companies, airlines, telephone line suppliers, estate agents and many more. When firms are few, and monitoring each other presents no great problem, firms can easily agree with no need of 'muscle'. The three top Italian producers of lead and amianthus pipes, for instance, are said to share territories – north, south and centre – when it comes to large orders. These firms do not seem to need any special agency to enforce their understanding. When it comes to small contracts, which are too arduous to police, they simply prefer to compete.

But other sectors – as diverse as construction, transport and street-hawkers – have been territorially controlled by mafiosi in Sicily. Enforcing the territorial allocation when markets are both visible and immobile (or at least well defined as bus routes are) is relatively simple. Territoriality has a logical basis that makes it the dominant alternative. In Naples and Palermo mafia enforcement has also helped prevent conflict among unofficial parking attendants, whereas in a similar market in Rome, that for cleaning windscreens at traffic lights, the mafia has not been available. In 1989 this business was lucrative: each man (mostly Polish immigrants) made between 100 000 and 300 000 lire per day ($80–250) and, according to the carabinieri, this led to violent fights over the allocation of the most profitable road junctions. The cleaners were unable to cooperate and thus attracted the attention of the authorities (*la Repubblica*, 2 July 1989).

4.2 Customers

Not all markets are efficiently allocated by territory. For instance, dealers may congregate in one location, as in wholesale markets, or agents may be highly mobile and unpredictable, as, for example, taxi drivers and, as noted by Schelling, burglars, who are both mobile and, as a matter of trade, invisible (Gambetta, 1993: 220–35). Furthermore, geographical divisions are not always satisfactory because profits may be so unevenly and unpredictably distributed (e.g. because of differences in regional patterns of development) that some cartel members perceive the continuation of the agreement as inequitable. An alternative solution consists in sharing customers: cartel members, in other words, agree not to accept or seek business from customers who are currently served by another member of the cartel. Where feasible, customer allocation dominates the other rules. Violations are easily detected within a short period of time and almost automatically because if a cartel member provides service to the customer of another member, that member will be aware of the loss of business of that customer. Moreover, selective sanctions can be applied by aggressive solicitation of the violator's customers.

The rule is also equitable and simple to apply. It gives to each participant what he thinks of as his already, namely his existing customers. It does not require other forms of cooperation which leave paper trails for investigators. Members can charge different prices and provide different qualities of products, though there may be limits on the variation depending on the ease of comparison by customers. This type of sharing is particularly attractive if customers are 'fixed in location and the service or good is delivered to [them]' (Reuter, 1987: 7) for policing is simpler. More generally, it is successful if the conditions permit an easy way to identify customers and keep track of their movements. When customers are big, for instance, they are easily shared between suppliers: A works for the electric company, B for the water company and so on.

This applies to the wholesale fruit and vegetable market too in Palermo. In fact, although most daily transactions now take place competitively, there is an exception which still provides mafiosi with an opportunity to exert their influence. A group of six or seven *commissionari* – those said to be of 'greater respect' – collude over the contracts taken out by institutions, such as schools, hospitals, barracks and old people's homes. Each *commissionario* supplies specific institutions and the problem of competition is eliminated radically: by common agreement only one *commissionario* shows up at each tender. Relative to private customers, public institutions are fewer, take out longer-term contracts, buy fixed quantities, are generally slack about quality, careless about price and corruptible. In short, they are easy to share. The small number of participants, as well as the availability of sanctions – they share the same crowded market area and could easily make life difficult for each other – are such that this agreement could in principle do without outside assistance, although some evidence suggests that this is not the case (Gambetta, 1993: 206–14).

Under customer allocation agreements, customers effectively become part of the supplier's assets; they are internalised and are in some instances traded like any other form of property. Carters in the New York metropolitan area (Long Island, New York City and New Jersey) buy and sell customers both individually and in groups. The prices of those customers, roughly 40 times their monthly gross billings in the early 1980s, provide the basis for estimating the capitalised value of the allocation agreement (Reuter, 1987: 48–51).

4.3   Queues

Sharing customers, however, can prove arduous. There may simply be too many of them around, or they may be too occasional or too mobile. In short, they can be difficult to identify: how could, say, restaurants share tourists or petrol stations share motorway drivers? Until the mid-1950s, when *commissionari* were very few at the fruit and vegetable market in Palermo, some ordinary customer sharing went on. Since both customers and sources of supply have increased, however, these arrangements have been abandoned. One strategy for dealing with these problematic markets involves taking turns. Provided customers can be funnelled through particular locations, queueing can help. The obvious case is that of taxis who line up at stations and airports.

Sharing customers is also meaningless if there is only one buyer, or, more generally, if there are fewer buyers than sellers. If there is just one buyer, queueing can work if purchases are repeated: sellers can then enter a (metaphorical) line and share the market on this basis. One common method of collusion found among building contractors in Sicily runs as follows: firms A1, A2, A3, ... agree that, say, A2 should obtain the contract; the others bid artificially high prices so that A2's offer is certain to be the lowest. There are certain prerequisites for a deal of this kind to work: first, only firms in the cartel can participate while free-riders must be excluded. Next, A2 must be confident that none of the other firms will submit a competitive bid at the last minute. In turn, A1, A3 ... require guarantees that A2 will return the favour on a future occasion and entrepreneurs therefore keep careful records of who obtained which contract. Where a contract is publicly bid on a repeat basis, the cartel may instead allocate a contract to one firm, with others designated to provide 'courtesy' bids as camouflage for the arrangement. That behaviour has been observed among the carters in both New Jersey and Long Island. These rights can be permanent enough for the contracts to be sold from one carter to another.

5 Conditions of Emergence

The ideal strategy from the entrepreneurs' point of view is to collude without paying protection money to the mafia, but this works only so long as everybody follows the rules and everybody believes it. Sometimes, racketeers can be called in simply *ex post facto* to provide a one-time service: once, in the late 1970s, according to a Sicilian building contractor, two firms suddenly defected and went after a forbidden contract so that 'the man of respect was called in and persuaded them to withdraw'. But, generally, not much is known about the conditions that foster the *continued* presence of the mafia in legitimate industries. In the United States the phenomenon of racketeer-influenced industries, despite a substantial journalistic interest, has attracted little scholarly attention. As for Sicily, research on this topic is virtually non-existent.

There seems to be just one known case, the waterfront industry in New York, in which the presence of racketeers qualifies as 'pure extortion'; and even so, employers take advantage of their presence to control union members (Reuter, 1987: vii; Bell, 1960). But existing studies – e.g. Landesco (1929), Block (1982), Reuter, Rubinstein and Wynn (1982), Reuter (1987) and Gambetta (1993) – found evidence that competitors seeking a collusive solution to market problems provide racketeers with the opportunity to acquire a role in their industry. Racketeers can enter 'by invitation' rather than on their own initiative. Recent evidence from Sicily illustrates this case. Baldassare di Maggio, a mafioso who turned state witness in 1992, said that Angelo Siino, a building contractor, 'came to me and said that if ... we, that is Cosa Nostra, were capable of coordinating the bids we could get much bigger profits. ... In that first stage, Cosa Nostra's problem was that of making [Siino] credible in the eyes of other contractors' (*Panorama*, 11 April 1993). In order for collusion to succeed, the organisational force of the cartel must be credible: every participant must respect the agreement and in turn confidently expect that his moment to bid will come and that others will respect it. Cosa Nostra supplies these guarantees. A surveyor involved in the same enquiry said that the mafia intervenes as 'an organization which takes care of the way in which the various jobs are equitably distributed among the interested firms' (*la Repubblica*, 28 October 1992). In the interrogation which originally started this enquiry in 1990, the former Mayor of Baucina, near Palermo, summing up the role of the mafia, said that the local mafiosi 'oversaw the fair distribution of contracts among firms participating in bids' (*la Repubblica*, 14 April 1990).

There is a parallel with New York: federal prosecutors in the case United States vs. Salerno in 1985 proved that 'the Cosa Nostra families established a club of concrete contractors who decided which contractor would submit the lowest bid on each project; other cartel members prepared their complementary bids accordingly. The "lowest bid" was far higher than the price that fair competition would have produced' (New York Organized Crime Task Force [OCTF], 1988: 11, fn. 17).

In general, racketeers offer the prospect that the conspiracy will work, simply because they provide credible enforcement. Large-number cartels appear generally to have short lives because there is always an incentive for a member to leave the agreement and take advantage of the restrictions imposed on the remaining members (Scherer, 1970: Ch. 6). Potential conspirators are aware of this and may be reluctant to enter into an arrangement that is probably short-lived as well as illegal. By promising to take illegal but effective actions against defectors, racketeers provide potential members with credible assurance that it is likely to be of lasting benefit.

But the racketeer involvement has another benefit, once it becomes known. It reduces the willingness of customers to protest the high prices charged under the agreement or to solicit competing bids from other dealers, tempting violation. Note that the customers in the commercial sector of the carting industry in New York include major corporations competent at understanding and exercising their rights. The unspoken threat, perhaps occasionally articulated in a vague way, that aggressive action may be punished by mafia interventions, smooths customer relations for the cartel members. Given the small share of total costs associated with garbage collection, and the fact that rivals are also believed to be extorted, risky resistance is unlikely to be an attractive option. The New Jersey State Commission of Investigation (1989) noted that one carter told a reporter that a story about racketeer involvement in the industry would only help his business.

Finally, racketeers provide a reputational barrier to entry. Entrants must be concerned that they will be the target of retaliation by racketeers. Their trucks and stores may be destroyed and their customers threatened or actually subject to violence. This was certainly the case when the

Brooklyn District Attorney's Office started a carting firm in 1972, as part of an undercover investigation of the industry.

Collusion alone, however, is not in itself a sufficient cause. There is no evidence of mafia control among large corporations – steel, automobiles, chemicals, rubber (Bell, 1960: 176; Reuter, 1987: 7). Entrepreneurs of this calibre can probably count on political collusion and, more generally, grander means than the mafia (e.g. Friedman, 1988). Nor does it develop in hi-tech industries. For markets to be vulnerable to the mafia, collusion must be both highly desirable and difficult to bring about. The former condition depends on inelastic demand and little product differentiation, whereas the latter is due to impediments such as numerous firms and low barriers to entry (Reuter, 1987: 6).

Even when all the right conditions obtain, however, a triggering event often seems necessary to attract racketeers. Slumps in the economy, for instance, can provide the activating force: in the United States during the Depression, an extraordinarily deep and rapid decline in demand intensified the incentive for collusive action and related services. The need for cash becomes an important factor: Lucky Luciano, a leading mafioso during the Depression, claimed:

[W]e gave the companies that worked with us the money to help them buyin' goods and all the stuff they needed to operate with. Then, if one of our manufacturers got into us for dough that he could not pay back, and the guy had what looked like a good business, then we would become his partner ... we actually kept a whole bunch of garment manufacturers alive, and we helped all them unions, the Ladies' Garment Workers and the Amalgamated, organize the place.

(Gosch and Hammer, 1975: 77–8)

Racketeers' intervention can also be sparked off by the sudden greed of one of the cartel participants. Funeral homes in Naples, for instance, shared hospitals for many years without outside intervention: firm A was responsible for the deceased in hospital X, B for those in hospital Y, and so on. They managed to keep the number of firms surprisingly low: while in Turin there are 50 firms for 1 million residents and in Rome 70 for 2.8 million, in Naples a population of 1.2 million is supplied by only 13 firms. The cartel achieved these impressive results with the assistance of corrupt local politicians and administrators who rejected new applicants. But collusive arrangements remain exposed to the risk of foundering and hence to the temptation to call in the mafiosi. At the end of the 1970s one firm tried to increase its share by enlisting the protection of an aggressive

Table 4.1   Conditions favouring the emergence of mafia-controlled cartels

Product differentiation	Low
Barriers to entry	Low
Technology	Low
Labour	Unskilled
Demand	Inelastic
Number of firms	Large
Size of firms	Small
Unionisation	Present

gang, La Nuova Camorra Organizzata. The other funeral homes attempted first to repel the move 'politically', then called in rival racketeers, who did not limit themselves to providing a one-time service but decided to continue levying a protection fee even after the triggering event had been sorted out. A similar case is found in the New Jersey garbage collection industry; there, however, the aggressive carter was not backed by outside racketeers and the other carters successfully opposed his attempts to expand by 'political' means alone.

According to one Sicilian contractor, mafioso intervention is much favoured by submissive expectations and can therefore be avoided; he himself had coordinated queues, ostensibly without being a mafioso. By yielding too quickly to the mafia – he claimed – one loses the respect which may otherwise be sufficient to control collusion: 'a lot of entrepreneurs end up in the arms of the mafia simply because they believe it is inevitable'. In New York the OCTF also found, in its investigation of the building industry, that 'sometimes contractors claim not to know exactly why they pay; experience tells them that payoffs are necessary to assure that "things run smoothly" ' (1988: 17). The stronger the rumour that mafia services are indispensable in a certain market or area, the higher seems the likelihood that they will be requested or accepted without question.[5] In addition, institutions looking for builders may themselves feel lost without an outsider to choose for them for no-one wants to take responsibility for the allocation: in at least one case the contracting agent in Sicily actively sought the intervention of an external fixer when the latter was slow in coming forward autonomously.

6   Consequences

Although it is difficult to establish *ex post facto* whether a given collusive arrangement could have emerged without the involvement of the mafiosi, the availability of mafia services generally makes collusion more likely, more elaborate and more enduring. Thus, not only do collusive arrangements sustain the mafia, but the availability of mafia protection also provides an incentive to seek collusive solutions.

Agreements supervised by mafiosi can embrace larger numbers of both dealers and customers: in the New York carting industry, for instance, in which racketeers played a continuing role in the operation of the allocation agreement, primarily through the constant need to mediate disputes, the cartel involved the allocation of over 100 000 customers to as many as 300 carters (Reuter, 1987: 11). The greater the number of firms the harder it becomes to ensure that they all keep their word. An Italian building contractor spoke of a 'queue' of as many as 160 firms. It took a great deal of patience and manoeuvring before he eventually succeeded, as he put it, in 'buying' a contract: 'it is obvious that to safeguard agreements of this size one needs the threat of violence'. This threat was not supplied by an entrepreneur belonging to the cartel; at most 'he owns a few caterpillars'. Without such a threat, as the OCTF in New York also found, 'bid-rigging conspiracies may founder because the conspirators are unable to police their cartels effectively' (1988: 38).

Irrespective of the method adopted, the general consequences of restrictive practices, as Reuter argued, are threefold: less efficient production, higher prices, and smaller firms.

Less efficient production is engendered by the reduced incentive for lowering production costs; a firm cannot obtain an increase in market share by lowering costs, since all existing customers are allocated. ...
The agreement also permits inefficient firms to stay in the market and prevents efficient firms from growing. The higher prices result directly from the imposition of restraints of trade. ... In each dimension, the effect is likely to be greater for a racket-run cartel than for other cartels.
(Reuter, 1988: 7)

The first victims of collusion are consumers who end up purchasing lower-quality goods for higher prices. Potential competitors come next: successful collusion makes it almost impossible for outsiders even to contemplate entering an industry, and overt intimidation becomes redundant. This is why the enforcement of internal agreements is much more important than the brutal discouragement of rivals, even though the lack of entrepreneurial energy in the south of Italy makes the function less important there. Finally, since racketeers increase the confidence of participating entrepreneurs that the cartel will endure, incentives for efficient production are even more sharply reduced than they would be in a conventional cartel, where certainty about future success is always limited and the probability of imminent competition never vanishes. The depressing effect on quality caused by mafia involvement can go further, especially in Sicily where inter-mafia conflicts have been more common than in the United States. The operation of the mafia is based on its reputation for effective contingent violence but that does not mean that actual violence is unknown. Both in New York and Sicily an occasional killing or act of violence has been required to discipline a cartel member. Though these acts are situation specific, they also serve to enhance the reputation of the mafia and augment the reputational barrier. However, when such acts of violence become 'too' frequent because of internecine wars – in Sicily during the 1980s several entrepreneurs died in the crossfire (Gambetta, 1993: 191–4) – the barrier becomes so strong that those very few who venture to take up entrepreneurial activities must be selected from risk-prone individuals who both have little chance of an alternative (and less dangerous) career and are already tightly connected with a mafia family. Even discounting for the lack of incentives for efficient production enjoyed by mafia-protected cartels, this group is most unlikely to contain characters versed in proper entrepreneurial tasks.

7   Conditions of Disappearance

Again we note that our set of case-studies is too small and diverse for the development of general propositions. The examples are interesting in themselves, however, as pointing to the variety of circumstances that can lead to the demise of mafia control. The fruit and vegetable market in Palermo is now largely competitive. Agents appear more concerned with quality than with establishing or violating collusive deals. The threat of violence seems as remote as in any normal business environment. Local authorities can take none of the credit for the market's evolution: 'The contacts which our group had with the present Mayor and the assessore all' 'Annona', concluded the Anti-Mafia Parliamentary Commission in 1969, 'gave us the impression, increasingly so, of lack of both interest and information, often coupled with open irritation towards those trying to unsettle a status quo which was not altogether disliked'.[6]

For years local authorities tolerated unfair dealing: for instance, an area of the market designated for use by producers selling without the intervention of middlemen was invaded by the most powerful *commissionari* who then cheekily levied a 10 per cent charge from farmers wanting to share it. In 1969 the Commission asked the Mayor why an area adjacent to the market was left unused instead of being developed into further stands. The Mayor replied that it was the site of a church that could not be demolished because of its artistic value. A letter of inquiry to the Soprintendenza delle Belle Arti of Palermo went unanswered. The members of the Commission went to see for themselves: they found a small chapel of no aesthetic merit being used as a garbage dump. In the same period, beautiful buildings, such as late nineteenth-century villas by Basile, were ruthlessly demolished in Palermo to make room for urban speculators (Chubb, 1982: Ch. 6).

It was thanks to the central authorities, the Prefects and the Anti-Mafia Commission, that the process of expansion, already to some extent 'naturally' under way, was accelerated. The present director of the market concedes that the Palermo Prefect's policy of increasing the availability of permits was crucial in alleviating the troubles of the market. Permits rose particularly during the 1970s until the last increment, in 1981, brought the number of licences to 77.

Yet, if the action of public authorities were the sole factor which brought about competition, why has the same process not occurred at the fish market? The answer lies in the diverse nature of the commodities involved. The supply of fresh fish is constrained both by limited natural resources – which are either unaffected or depleted by technological development – and by the fact that it is channelled through a limited number of clearly identified ports. Fishing firms, furthermore, are fewer and easier to control than farmers. In the fruit and vegetable trade the sources of supply are both many and widely scattered, and transport routes and delivery points outside the wholesale market can multiply unnoticed.

The rapid increase in agricultural productivity and conservation techniques after the war, as well as the improved efficiency of transport, accentuated these differences and made it virtually impossible to monopolise produce. The only example of collusion mentioned by the Commission involving *commissionari* and their emissaries concerns a case in which distribution is forced through a bottleneck: by controlling the boats linking the small island of Pantelleria to the main island, a group known as L'Associazione made sure that the local cooperative sold the zibibbo grape exclusively to them. But, in general, monopolies were already being eroded in the 1960s because, according to the Parliamentary Commission, 'wealthy wholesale and retail dealers collect directly from northern Italy and introduce ever-increasing quantities of good fruit for consumption'.

A further bottleneck has also been broken. Before the rural Casse di Risparmio started to function properly, credit was in the hands of the 'man of respect'. Producers who wanted credit were obliged to sell their goods to him. As sources multiplied credit no longer had the same kind of strings attached. It is now granted sparingly to customers rather than producers and is a means of fostering competition rather than enforcing collusion. Creditors are less inclined to resort to violence against insolvent debtors lest other clients are diverted to more benign rivals.

There are also instances of the New York mafia losing its influence in industries. The trucking industry, in which the mafia notoriously had acquired influence throughout much of the nation through the instrumentality of the Teamsters Union (three of whose presidents have been convicted of corruption), now seems to be much less subject to that influence. Two factors may have been significant here; deregulation of the industry (which reduced the power of the union) and aggressive and imaginative prosecution of the Teamster leadership (which resulted in a reform leadership being installed at the beginning of the 1990s).

However, most targeted efforts to remove the mafia from specific industries in the United States have been generally unsuccessful. In response to concerns about the role of the mafia (and lack of competition) in the carting industry in New Jersey and New York City, the industry has been subject to regulation (since 1956 in New York City and since 1968 in New Jersey). A flow of cases shows that the regulation has been utterly ineffective in New York City; the evidence concerning New Jersey is less clear but anti-competitive agreements, perhaps with mafia involvement, have been alleged as late as 1988. The regulatory apparatus had indeed become one of the tools for the operation of the customer allocation agreement. The regulators allowed the sale of customers and, through inflexible price rules in New Jersey, provided the basis for contesting new entry.

8   Prices and Profits

The returns to participants in these cartels are difficult to determine. The available data are best for the carting industry, where customers are treated as assets and frequently bought and sold by the participants (Reuter, 1987).

The little we know of the charges by mafia members for their services suggests that, at least where the relationship with the industry is stable, those charges are surprisingly modest, indeed well below extortionate levels. In the Long Island carting industry in the mid-1980s it appeared that the mafia took no more than $400 000 annually in tribute, though the estimated profits accruing to the carters was over $10 million. The mafia-run concrete cartel in New York City levied only 2 per cent of the contract price for its services in fixing prices. As for Sicily we have the following evidence: in 1993, Baldassare di Maggio, a *pentito* we encountered already, describing the mafia's role in organising bid-rigging in construction, reported that they kept 5 per cent; 3 per cent was for the mafioso organisation and 2 per cent was given to Siino to pay the politicians that were paid as part of the overall agreement.

The fact that customers of carters are sold at such high prices in inter-carter transactions provided evidence that the primary beneficiaries of the agreement in that industry were the carters themselves. If the mafia taxed away all the returns from the agreement, then customers would have no greater value than they would in a competitive industry; in fact they sold for a very substantial mark-up over the competitive price.

We can offer no convincing explanation for these observations. If the mafia were marketing its cartel-organising services, then these might be seen as prices intended to maximise their profits over the entire market for such services. However, the peculiar lack of mafia entrepreneurialism – no new industries have been identified as being subject to mafia control over the last decades – weakens this supposition. Nor does it seem that the mafia lacks information about the profits generated to other actors in this business. Access to accounts of member enterprises is certainly not a problem in the New York markets in which the organisation is active. Finally, we do not believe that these prices are set so as to minimise the risk of aggressive informing by other participants.

9 Conclusions

We make no claim to have identified all the relationships between the mafia and legitimate enterprises. The casino industry in Nevada, where the mafia interest was confined to tax fraud, debt collection from bettors, and some backruptcy scams involving capital from the Central States Teamsters pension fund, represents a very different situation. On the waterfront, in New York and Florida, the mafia seems to have used its power, again through the instrumentality of a union, to extort shippers, carriers and those providing services on the docks. In neither of these industries has it acted to suppress competition on a continuing basis. Yet, we believe that the phenomenon described in this chapter is one that has received too little attention. The suppression of competition is a near-universal dream of established entrepreneurs. The mafia is one of the few non-governmental institutions that can help accomplish this goal. Our impression, based on a limited set of examples, is that once the mafia has entered, its success in meeting the needs of its customers, as well as its unique reputational asset, makes the reintroduction of competition extremely difficult. The phenomenon observed here may be just a generalised form of the Stigler–Peltzman theory of state regulation (Stigler, 1971; Peltzman, 1976). The state is unable or unwilling to meet the demand of entrepreneurs for profit-enhancing and risk-reducing regulation; the mafia provides the service instead.

The phenomenon is perhaps of more analytical than policy interest, since there are signs that it is now receding, mainly because the mafia itself appears to be diminished in power in both countries. In the United States the results of a continued federal campaign have led to the conviction and long-term imprisonment of much of the leadership. In Sicily too the mafia never had it so bad and is now under unprecedented state pressure: virtually all mafiosi who sat in the mafia 'governing body', known as La Commissione, are now in jail. Mafioso assets worth 3500 billion lire were seized by customs police in 1992 alone. At the same time a new crop of mafiosi turning state witnesses has been growing and their revelations are leading to further arrests and repressive actions.[7] Mafia reputation may be, initially at least, little diminished by these imprisonments, but its ability to venture into new fields is probably limited by the inexperience of, and instability in, the new leadership. Furthermore, the current dramatic political changes in Italy are likely to decrease the opportunities for political corruption and bid-rigging which used to provide the major outlet for mafia-regulated cartels of building contractors. Finally, some of the market conditions that (jointly) facilitate mafia entry as a cartel organiser - such as low technology, unskilled labour and unionisation – occur less frequently now than when it acquired its influence over industries, mostly between 1920 and 1950.

The prospects for increasing our understanding of the phenomenon are being enhanced by the very forces that are diminishing the mafia's powers. Certainly the trials of the major families in both Sicily and various American cities (including Boston, Chicago, Kansas City and New York) have provided a wealth of detailed information on the relationship between the mafia and their legitimate clients which awaits analysis.

Notes

1 Here we prudently use the term mafia as referring to the Sicilian mafia, also known as Cosa Nostra, both in Italy and the United States. Whether other broad and enduring criminal organisations supply the same service is a matter for empirical analysis to establish.

2 A curious case of failure of cooperation due to difficulties of both policing and sanctioning is represented by radio taxis in Palermo (see Gambetta, 1993: 220–5).

3 Carting is the term used to describe the collection of solid waste, whether from households or commercial establishments. It does not include the collection of hazardous waste, where different possibilities exist for collusion and fraud.

4 'Up to perhaps six firms one has oligopoly, and with fifty or more firms of roughly similar size one has competition; however, for sizes in between it may be difficult to say.' This is the rule of thumb offered by James Friedman's textbook (1983: 8).

5 This is analogous to the argument which holds that widespread corruption is self-generating: the expectation suffices to produce more of the thing itself (Andvig and Moene, 1990).

6 The case of this market is illustrated in full in Gambetta (1993: 206–14).

7 For a full assessment of the mafia's prospects in Sicily see the new introduction to the second Italian edition of Gambetta's *The Sicilian Mafia*, published by Einaudi (Torino) in 1994.

References

Andvig, J. C. and Moene, K. O. (1990) 'How Corruption May Corrupt', *Journal of Economic Behavior and Organization*, 13, pp. 63–76.

Bell, D. (1960) 'The Racket-ridden Longshoremen', in D. Bell (ed.), *The End of Ideology* (Glencoe, IL: Free Press).

Block, A. (1982) *East Side–West Side: Organizing Crime in New York, 1930–1950* (New Brunswick, NJ: Transaction Press).

Chubb, J. (1982) *Patronage, Poverty and Power in Southern Italy: A Tale of Two Cities* (Cambridge: Cambridge University Press).

Fiorentini, G. (1994) 'Cartels Run by Criminal Organizations and Market Contestability'. Unpublished paper, Department of Economics, Università di Firenze.

Franchetti, L. [1876] (1974) 'Condizione politiche ed amministrative della Sicilia', vol. 1 of L. Franchetti and S. Sonnino (eds), *Inchiesta in Sicilia* (Firenze: Vallecchi).

Friedman, A. (1988) *Agnelli and the Network of Italian Power* (London: Mandarin Paperback).

Friedman, J. (1983) *Oligopoly Theory* (Cambridge: Cambridge University Press).

Gambetta, D. (ed.) (1988) *Trust: Making and Breaking Cooperative Relations* (Oxford: Basil Blackwell).

— (1993) *The Sicilian Mafia: The Business of Private Protection* (Cambridge, MA: Harvard University Press).

Gosch, M. A. and Hammer, R. (1975) *The Last Testament of Lucky Luciano* (London: Macmillan).

Landesco, J. [1929] (1968) 'Illinois Crime Survey, Part III'. Reprinted in *Organized Crime in Chicago* (with introduction by Mark Heller) (Chicago: University of Chicago Press).

New Jersey State Commission of Investigation (1989) 'Solid Waste Regulation', Trenton, NJ.

New York Organized Crime Task Force (1988) *Corruption and Racketeering in the Construction Industry* (New York: ILR Press).

Peltzman, S. (1976) 'Toward a More General Theory of Economic Regulation' *Journal of Law and Economics*, 19, pp. 211–40.

Reuter, P. (1983) *Disorganized Crime: The Economics of the Visible Hand* (Cambridge, MA: MIT Press).

— (1987) *Racketeering in Legitimate Industries: A Study in the Economics of Intimidation* (Santa Monica, CA: Rand Corporation).

— (1993) 'The Commercial Cartage Industry in New York', in A. Reiss and M. Tonry (eds), *Beyond the Law: Corrupt Organizations*, vol. 18 of *Crime and Justice: A Review of Research* (Chicago: University of Chicago Press).

— Rubinstein, J. and Wynn, S. (1982) *Racketeering in Legitimate Industries: Two Case Studies* (Washington, DC: National Institute of Justice).

Scherer, F. M. (1970) *Industrial Market Structure and Economic Performance* (Chicago: Rand-McNally).

Stigler, G. J. (1964) 'A Theory of Oligopoly', *Journal of Political Economy*, 72, pp. 367–83.

— (1971) 'The Theory of Economic Regulation', *Bell Journal of Economics and Management Science*, 2.

5

Editors' Introduction

Gambetta and Reuter's analysis puts somewhat of an ironic twist on the notion of 'crime prevention' with regard to organised crime; it appears that the ostensible 'victims' may not be unwilling, and actual victimisation is diffused across the broad community of consumers who pay more than they would otherwise need to. It is fortunate that not all crime is muddied by such complexities, but prevention of 'ordinary' crime still provides an analytic challenge which the social sciences – like government and law enforcement – are only just beginning to address.

Architectural theories of defensible space, and social theories of situational crime prevention and multi-agency response, shift criminological attention from the deep background of criminal motivation to the circumstances more immediately surrounding decisions to commit crime. Debate has, of course, raged over the adequacy of understandings based on purely rational decision-making, perharps echoing criminology's roots in a nineteenth-century framework which saw criminal offending as the ultimate sign of irrationality. Equally, crime prevention programmes founded on anticipating the likely actions of potential offenders cannot do other than assume a logic which can be apprehended and applied to deterrence.

Criminological inquiry into the aetiology of crime has had modest practical returns, yet the implications of these analyses are not entirely marginalised by a perspective based on the economist's understanding of the offender as a choice-making social actor weighing the costs and benefits of given courses of action, as is apparent in Farrell, Chamard, Clark and Pease's model, which seeks to integrate criminological and crime prevention theory within the framework of economic analysis. Key variables are guilt and shame experienced by the offender, as well as measures of time, effort, risk and rewards. The model uses the formal economic framework to capture how decisions are made to commit crimes, to show why such decisions result in repeat victimisation of the same targets, and how an understanding of the logic of these decisions can be used to reduce offending.

Towards an Economic Approach to Crime and Prevention

Graham Farrell, Sharon Chamard, Ken Clark and Ken Pease

I Introduction

When deciding to commit an offence, offenders take into account different factors. The key factors or variables in the economic model developed in this chapter are time and effort, risk, rewards, and available excuses as perceived by the offender. This is in keeping with the rational choice model of criminal decision-making (Cornish and Clarke, 1986, 1987) which is situated within the framework of situational crime prevention (see Clarke and Homel, 1997). The criminological theory is not reviewed at length, since the focus is upon transforming it into an economic framework. The principal influence in the criminological literature addressing crime and victimisation in an economic-type framework is van Dijk's (1992) macroeconomic model of the interaction between the rational choices of offenders and victims. This chapter takes a more micro-level approach, and shows how the model applies to the prevention of repeat victimisation. The contribution of Field and Hope's (1989) pioneering effort should also be acknowledged. It is intended that the present work should prove entirely compatible with these models.

Section II outlines the model of offender decision-making. Section III discusses how crime prevention works, that is, how the variables of time and effort, risk, rewards, and the removal of excuses, produce variations in the levels of crime. Section IV examines repeat victimisation and discusses the role of uncertainty in offender decision-making, and Section V discusses the possibilities for further research and development of the model.

The purpose of this chapter is not to develop criminological or crime prevention theory. Neither does it seek to advance the framework of economic analysis. This has been well developed since the time of Becker (1968). Our main aim is to begin to integrate the two; we contend that this synergy is the contribution. Specifically, the formal economic framework allows one to view clearly how offending decisions are made, how such decisions result in the same targets being repeatedly victimised, and how crime prevention interventions can reduce offending.

II The Model

It is assumed that people are utility maximisers. People decide to undertake an activity, including the commission of a crime, if one essential criterion is met. An activity will be performed if the perceived or expected potential net benefits from time spent in that activity outweigh the expected net benefit from time spent in any of the perceived available alternatives. In the same set of circumstances, different people will make different decisions depending upon a multitude of factors such as attributes, skills and experience. The decision to offend is not unlike an economic labour supply problem involving a time-allocation decision. Is it rational to spend time committing this offence rather than doing something else?

A brief note is necessary here on the notion of limited rationality, or what economists call 'bounded' rationality. Clearly, the decision made by an offender is not always correct. A wrong decision can be made, but it was the perceived optimal decision at that time and in those circumstances, from the viewpoint of the offender. Also, 'rational' decision-making does not refer solely to hardened offenders seeking suitable targets. Even a normally law-abiding person may break the law if the benefits of committing the crime are perceived to outweigh the benefits of not committing the crime.

A person's overall utility is derived from the time spent offending and time spent in other activities. Hence,

$$U = U(t_L, t_C) \tag{1}$$

where U is utility, t_L is time spent in legitimate activity (work and leisure), and t_C is time spent on criminal activity. Note that $t_C + t_L = T$ is the total time available for allocation. In any given period, most people will probably spend all of their time on legitimate activity and none on criminal activity. The amount of time spent on criminal activities will vary with the particular characteristics and experiences of the person as well as circumstances. From a person's particular set of physical and mental attributes, skills and experience, is derived their particular preference curve for committing a particular criminal act.

The relationship between the economic concept of utility and the variables mentioned in the offender decision-making process is relatively straightforward. The individual's decision is based on whether the additional benefit of spending an extra unit of time in criminal activity is outweighed by the cost. This additional benefit is the marginal utility of criminal activity and is related to the four factors of the situational crime prevention approach. These are (1) expected cost due to the perceived time and effort spent on offending, (2) expected cost due to risk, (3) the expected psychological cost of guilt and shame (when excuses are absent), and (4) the expected benefit (whether monetary or psychological) of the rewards. Rewards are the only benefit perceived by the offender, and for a crime to take place they must be perceived to outweigh all of the costs. The marginal utility of criminal activity is expressed as a function of time and effort (e), risks (k), the absence of excuses (G), and rewards (R).

$$U'_C(e, k, G, R) \tag{2}$$

As effort or risk increases, or as rewards or the level of available excuses decreases, the marginal utility of crime decreases. From the perspective of most people in most instances, the actual or perceived costs will far outweigh the rewards, and so they do not commit crime.

If an individual believes the expected returns from spending time committing crime are greater than those expected from non-criminal activity, then more time will be allocated to criminal activity. It can be safely assumed that at some point the rewards per minute of time allocated to offending will begin to diminish. The optimal time-allocation point is reached when extra time spent offending would cost more than not offending. The amount of time spent offending is derived when the marginal utility of crime, U'_C, is equal to the marginal utility of non-criminal activity, U'_L, so that

$$U'_C = U'_L \tag{3}$$

The intersection of the marginal utility curves of time spent offending (U'_C) and time spent in non-offending activity (U'_L) is shown as Fig. 5.1a. Total time is shown on the horizontal axis, and runs from zero to the maximum, T. This total is broken down into two components: time spent on crime, which runs between zero and t_C^*, and time spent on non-criminal activity, which is measured by the distance between t_C^* and T.

Different people have different sets of preferences, which means they have different marginal utility curves for crime. For example, even if some people are presented with a criminal opportunity, they prefer not to take it. This is because they do not perceive a net benefit when factors such as guilt and shame are incorporated. Such a situation is shown in Fig. 5.1b as the curve U'_{C2} which always intersects U'_L at a point where zero time is

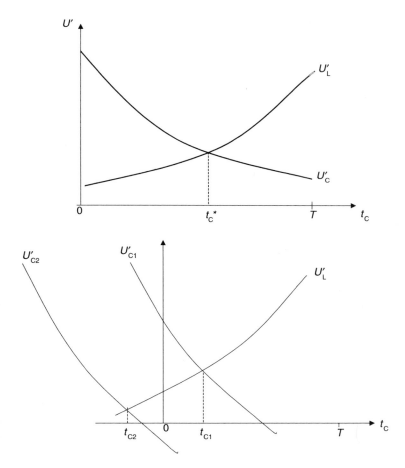

Figure 5.1 (a) Allocating time to crime: the marginal utility curves of time spent offending and time spent not offending; (b) a preference for crime? Different marginal utility curves results in offenders and non-offenders

allocated to crime. This contrasts with the person with a set of preferences that result in the marginal utility of crime curve of U'_{C1}. This intersects with the marginal utility of non-criminal activity curve, U'_L, at a point where the amount of time between T and t_{C1} is allocated to criminal activity.

The amount of time a person allocates to commit crime will vary greatly with the crime type, circumstances, and the particular person. Many variables influence the relationship between time and crime. Different crimes take different amounts of time to commit, with wide variation by location and circumstance, people and their abilities.

The next issue is the fact that time and effort are related. Other things being equal, making something harder to do (such as making crime harder to commit), requires more time. The model of crime as a function of time is already, in effect, a model of crime as a function of time and effort. More time is more effort, and vice-versa. This is important for what comes later since many crime prevention interventions make crime more difficult, requiring more effort. This translates into the model as requiring more time for the commission of the offence. Any variations on the assumption that time is effort, such as notions of a degree of exertion, can more readily be introduced at a later date once the basic model has been established.

Next comes the fact that time and risk are inherently related. Risk is, as with most economic models to date, a function of the risk of arrest, conviction and punishment, where risk is increased primarily by increasing the severity of punishment. In the present model, if it takes more time to commit a crime, then assuming risks are constant per unit of time, risks increase proportionately with time. For example, car steering-wheel locks prevent crime by deterrence since it takes more time and effort to remove the lock, during which time someone may observe the crime and mobilise a capable guardian. Hence the broader definition of risk, denoted by K, is a function of time and effort, e, and other risks, k

$$K = K(e, k) \tag{4}$$

While the basic model of crime and prevention is a model of risk, the determinants of risk are broader than commonly specified in most economic analyses of crime. Time and effort carry far more weight as components of risk. They are explicitly rather than implicitly incorporated, and play a fundamental role. In the following section, when the techniques of situational crime prevention are presented, the roles of the different components of risk should become clearer because different elements of risk may be manipulated via different crime prevention interventions.

In addition to the changed definition of risk, the present model emphasises both rewards and shame. These can be directly influenced by crime prevention policy, and are significant influences upon the decision to offend. In previous models they are typically viewed as fixed factors or givens rather than variables that can be manipulated. Again, the role of each of these concepts is explained more fully in the following section. They are mentioned here in order to help clarify how the present economic analysis is intended to be different from previous ones.

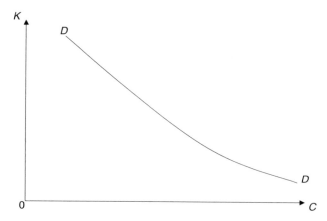

Figure 5.2 Risk and crime model

Drawing the elements together into Fig. 5.2, the quantity of crimes committed, C, is shown on the horizontal axis, with risk, K, on the vertical axis. Note that this single relationship now incorporates each of the individual components of equation (2). Risk, K, now incorporates time and effort (t) as well as other risks (r). Similarly, as will be shown below, the curve of the demand for offences by offenders, DD, can be altered by changes in perceived rewards or the level of available excuses. The risk – crime relationship is assumed to be negative since, as it becomes more difficult to commit crime, the level of crime people are willing to commit will fall.

In what follows, the demand for offences by offenders is met by a supply of criminal opportunities. In keeping with van Dijk's terminology, victims (and potential victims) are the reluctant suppliers of criminal opportunities (van Dijk, 1992: 106). At any given point in time there is a given level of criminal opportunities in society. This may change over time as, say, the volume of consumer goods available for theft increases. Changes in the supply of criminal opportunities, perhaps the result of public policy or private interventions, play an important role in the present model. Although the supply of criminal opportunities was wonderfully developed in van Dijk's 1992 model, it is a factor that is often assumed to be constant in economic analyses of offender decision-making and crime policy.

III   How Does Crime Prevention Work?

Crime prevention interventions increase the expected time and effort required to commit an offence, increase the expected risks of detection (and perhaps arrest, conviction and punishment), reduce the level of excuses that enable the commission of an offence, or reduce the expected potential rewards. This is the current model of situational crime prevention as proposed by Clarke (1997), based upon Clarke and Homel (1997). The table of 16 techniques is reproduced as Table 5.1, and provides concrete examples of prevention techniques, giving a real-world backdrop for the present chapter.

A   Increasing Expected Time and Effort

As detailed in the previous section, increasing the time and effort required to commit an offence will result in an increased risk of detection. Detection involves a subsequent risk of apprehension and punishment, whether formally or informally meted out. Table 5.1 shows specific examples of how time and effort can be increased. The most popular example is target hardening, although the spectrum of possibilities is wide.

The previous section discussed why effort is time, and time is risk. Making a crime harder to commit means an offender must take more time (*ceteris paribus*) to commit it, resulting in more risk of detection. This may result in the offender being deterred. In the model, at any given level of risk, an increase in the time and effort required to commit an offence results in a reduction in the supply of criminal opportunities. While the demand for offences, shown as D_1D_1 in Fig. 5.3, remains the same, the supply of criminal opportunities curve shifts to the left, representing a reduction in supply. The increase in risk from p_1 to p_2 produces a decrease in offending from c_1 to c_2.

B   Increasing Expected Risks of Detection

Increased risk of detection during the commission of a criminal act could result in a range of outcomes. Conceivably, a person detecting the offence may take no action if that person is a disinterested passer-by, someone who 'does not want to get involved', or someone incapable of taking action against the offender. Detection in this model therefore assumes detection by a capable guardian. This can result in a range of formal or informal outcomes, with the risk of arrest, conviction and punishment at the more formal end of the spectrum. The primary way risk is manipulated in the present model (excluding time and effort outlined above), is not via

Table 5.1 Sixteen techniques of situational crime prevention

Increasing perceived effort	Increasing perceived risks	Reducing anticipated rewards	Removing excuses
1. *Target hardening* Slug rejecter devices Steering locks Bandit screens	5. *Entry/exit screening* Automatic ticket gates Baggage screening Merchandise tags	9. *Target removal* Removable car radio Women's refuges Phonecards	13. *Rule setting* Harassment codes Customs declaration Hotel registrations
2. *Access control* Parking-lot barriers Fenced yards Entry phones	6. *Formal surveillance* Red-light cameras Burglar alarms Security guards	10. *Identifying property* Property marking Vehicle licensing Cattle branding	14. *Stimulating conscience:* Roadside speedometers 'Shoplifting is stealing' 'Idiots drink and drive'
3. *Deflecting offenders* Bus stop placement Tavern location Street closures	7. *Surveillance by employees* Pay-phone locations Parking attendants CCTV systems	11. *Reducing temptation* Gender-neutral phone lists Off-street parking	15. *Controlling disinhibitors* Drinking-age laws Ignition interlock V-chip
4. *Controlling facilitators* Credit-card photos Gun controls Caller ID	8. *Natural surveillance* Defensible space Street lighting Cab driver ID	12. *Denying benefits* Ink merchandise tags PIN for car radios Graffiti cleaning	16. *Facilitating compliance* Easy library checkout Public lavatories Trash bins

Source: Reproduced from Clarke (1997), which was adapted from Clarke and Homel (1997).

increased punishment as with many economic analyses, but via increased risk of detection. For present purposes the level of punishment is an assumed constant that is outside the scope of influence of most people or businesses. On the other hand, increasing risks of detection is increasingly seen as a far more variable private and public policy crime prevention instrument than previously envisaged. The potential for manipulating indirect risks of being observed, by passers-by, other citizens, colleagues or persons working in an area who may serve as capable guardians, is gaining increasing attention in criminology. Such risks are increased primarily by activities that are not related to the police but threaten the mobilisation of the police or private security. In practice, the threat of detection often results in crime being deterred because the criminal opportunity has effectively been removed.

In terms of the model, increasing detection risks due to exit screening, formal, informal or natural surveillance (see Table 5.1), operates in a similar manner to increasing time and effort. Increased risk of detection produces a reduction (a leftward shift) in the supply of criminal opportunities, from $S_1 S_1$ to $S_2 S_2$ in Fig. 5.3, with risk increasing from p_1 to p_2, leading to a fall in the volume of crime committed from c_1 to c_2.

Clearly, there is a connection between increasing the level of expected risks of detection and increasing the level of expected time and effort required. Overcoming increased risks of detection can involve more time and effort. While they are portrayed as operating via the same risk

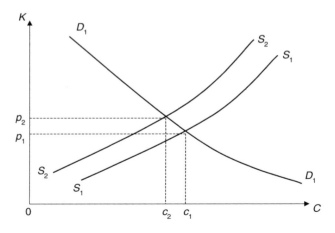

Figure 5.3   Increasing (time and effort and) risks of detection reduces criminal opportunities

mechanism in the model, in practice these interventions can vary greatly in nature, as shown in Table 5.1. Both serve to reduce the supply of criminal opportunities, but how this happens in the real world varies greatly.

It may appear that the present model loses details by integrating time and effort plus other risks into a single 'risk of detection' variable. However, the model helps demonstrate how the different techniques reduce crime by shifting the supply of criminal opportunities. The significance of this will become more evident as the operational mechanisms for changed rewards or excuses are described.

C Reducing Expected Rewards

If there is a reduction in the expected rewards to the commission of a crime then, at a given level of detection risk and criminal opportunities, fewer offenders will choose to commit the crime. At the extreme, removing the rewards will result in none of this type of crime being committed. Such changes are rare but might be said to have largely occurred in relation to thefts of work animals. When many work animals became outmoded due to automotive transportation, their re-sale value and hence profitability dropped, so that the market for theft of work animals diminished. A more modern example is the removal of subway graffiti in New York. Daily cleaning of subway cars meant that graffiti artists could not see their graffiti around the city. This removed the incentive or reward from the activity, and ultimately resulted in a large decline in graffiti (Sloan-Howitt and Kelling, 1997).

An actual or expected reduction in rewards leads to a drop in the perceived utility that an offender will derive from a crime. Looking back to Fig. 5.1b, a reduced reward might shift the preferences of an offender from curve U'_{C1} to U'_{C2} causing the offender to no longer wish to commit that offence. In the aggregate this lower level of willingness to commit a crime at a given level of risk and criminal opportunity manifests itself as a shift in the demand for offences. In Fig. 5.4 the original demand for offences by offenders is shown by D_1D_1, producing a level of crime c_1 at risk level p_1. With a reduction in rewards of a given offence, the supply of criminal opportunities does not change; instead the demand for offences shifts from D_1D_1 to D_2D_2. This produces a reduction in the overall level of offences from c_1 to c_2. For the lower level of rewards now available, offenders will be less willing to incur risk (time and effort and other risk of detection), so that the market level of risk they will incur at that level of criminal opportunities drops from p_1 to p_2.

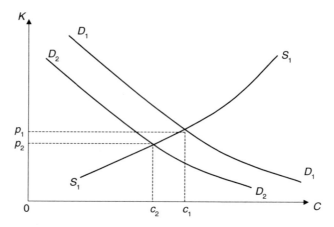

Figure 5.4   Reducing rewards or removing excuses reduces the demand for offences by offenders

D   Removing Excuses

Removing excuses is the most recent arm of situational crime prevention techniques. Setting rules, such as customs declarations that remove any excuse for a failure to declare taxable goods, removes the possibility of excusing oneself due to ignorance. Other simple measures, such as rubbish bins that facilitate disposal without littering, are among the range of tactics shown in Table 5.1.

In terms of the model, removing excuses operates via a mechanism similar to a reduction in rewards. Due to the lower level of excuses available, the utility of the offender has been reduced for the same level of risks and rewards. As illustrated in Figure 5.4, this shift in preferences results in an inward shift in the demand for offences curve from D_1D_1 to D_2D_2, with a drop in the level of offences committed from c_1 to c_2.

E   Combined Techniques

Four separate mechanisms for preventing crime have been presented in terms of the economic model. Clearly they need not, and do not, work in isolation. What if, for instance, an office manager makes computers harder for thieves to remove by fastening them to a surface, increases the formal surveillance of the office by locating secretarial staff at the entrance to meet newcomers, and also property marks the computers to reduce their re-sale value? The first aspect would increase the time and effort required to remove a computer (increasing risk); the second aspect would increase risks of detection (probably acting as a deterrent); and the third would reduce the rewards of stealing the computers since property-marked products are difficult to fence. The anticipated net reduction in crime would be the result of shifts in both the supply of criminal opportunities and the demand for crime by offenders.

IV Repeat Victimisation and Certainty

Once victimised, the same targets are more likely to be repeatedly victimised. There is an increasing amount of research examining how, where, why and when this occurs, and developing the implications for crime prevention and other areas of public policy (Pease, 1998). How would repeat victimisation be the result of rational offender decision-making in the context of the present model? Where the same offender is involved, the answer may be that repeat victimisation is rational since the level of uncertainty involved has been reduced.

Offending decisions take place under uncertainty. The offender seeks to minimise uncertainty by developing rational expectations of the costs and benefits related to an offence. Where there is uncertainty, at least one of the variables (risk, rewards, the level of excuses) is imperfectly known. That is, the actual costs and benefits are unknown quantities. Rather, the offender has estimates of what is expected. These estimates will cover a range of possible values, the range being narrower where uncertainty is least. Where there is great uncertainty relating to risk or rewards, for example the possibility of the costs outweighing the benefits, then *ceteris paribus*, the offender may be less likely to commit this offence. The levels of costs and benefits can be envisaged as estimated within parameters. Offenders may collect information to try to narrow the parameters by improving their knowledge of likely risks and rewards, such as when a burglar 'cases' a property. After committing an offence, an offender has far more accurate knowledge of the costs and benefits of that particular crime. When considering further offending, if an offender assumes that little has changed in the interim to make the target less attractive (which could be an incorrect assumption), there is less uncertainty involved in victimising that target again. The costs and benefits are known with greater accuracy unless the situation relating to the target has been altered. Hence, in some instances, repeatedly victimising the same target is a rational choice since it is known to be a more attractive prospect than the available alternatives.

In Figure 5.5, the offender does not know the exact risk that would be involved in committing a crime. It could be at the level p_1 or at p_2, or somewhere in between. In the aggregate, the number of crimes

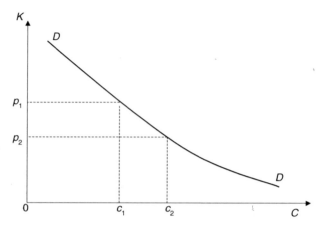

Figure 5.5   Repeat victimisation committed with reduced risk uncertainty

committed under such conditions might be about midway between c_1 and c_2. If, after committing the crime, an offender learns that the true level of risk is close to p_2, then, in the aggregate, the level of repeat victimisations committed will be close to c_2.

While this discussion of repeat victimization involved only uncertainty relating to risks, a similar uncertainty can exist in relation to rewards, or even the level of excuses. Since both operate via shifts in the demand curve, it is simpler for present purposes to talk of uncertainty relating to rewards. Fig. 5.6 shows offending under uncertainty in relation to both risks and rewards. Under uncertainty, offenders do not know if the level of risk is at p_1, p_2, or somewhere in between. Similarly, that offender does not know if the level of rewards would leave him/her on the demand curve D_1D_1 or D_2D_2 or somewhere in-between. Assuming a normal distribution of risk aversion, then the level of crime committed would be about midway between c_1 and c_2. If, on the other hand, offenders knew that the true level of risk was p_2 and that the true level of rewards would result in a demand D_1D_1, then the level of crimes committed would be far higher at c_2. Hence to keep crime at a minimum, creating the appearance of a greater than actual risk or lower than actual reward will result in a lower level of crime. Allowing offenders to estimate risks and rewards with certainty is a means to facilitate more accurate decision-making on their part. Note that in order to reduce or prevent crime, the role of uncertainty should be solely to make risks appear higher and rewards lower than in actuality.

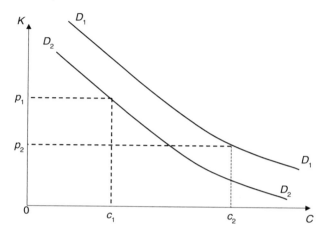

Figure 5.6 Offending under uncertainty of both risk and rewards

V Discussion and Concluding Remarks

The model presented may provide a platform upon which a range of theories and empirical studies can be developed. The model can easily demonstrate how 'hardened' offenders with a steep demand curve may be less responsive to reductions in opportunity than more amateur and opportunistic offenders with less steep demand curves. It may prove possible to model how organised crime reacts to prevention, particularly for crimes with exceptionally high profits such as drug trafficking. Crime rate changes due to shifts in routine-activity-type variables, and the manner in which they influence crime, are relatively easily shown. For example, the supply of criminal opportunities increases if there is an increase in consumer goods available, increasing the number of criminal opportunities at any given level of risk and thereby expanding the market for crime (as in van Dijk's 1992 macro-model). The whole area of the elasticity of demand with respect to changes in supply for particular crime types, their variation across time and space, and notions of cross-crime type (product) switching may relate to aspects of crime deflection, displacement and prevention in ways that have not yet been explored. It may also be possible to determine when the slope of the demand curve changes, which corresponds with the point at which people's routine preventive precautions require additional expenditure so that they become more expensive 'special' precautions. Crime prevention resources might be allocated more efficiently at those times and to those places, in relation to those crime types and circumstances, where the elasticity of demand for crime is greatest.

In many of the instances discussed above, the formal economic model may provide a clear framework for analysis, with the aim of informing policy development, as well as the potential for the development of empirical studies. Although a model of crime and prevention may never be as influential as Becker's model of crime and punishment, it is not impossible that, with further development and research, it could be used to inform optimal policy-making in relation to public and private crime prevention policy, allocations and expenditures. For example, it should be possible to model how and why crime prevention should seek to maximise uncertainty on the part of offenders in order to reduce repeat victimisation. Similarly, integrating the different perspectives into a comprehensive criminal justice model within an economic framework could be a long-term goal from which fruitful insights can be gained.

If the present model of criminal decision-making appears both simple and somewhat similar to previous economic analyses of crime, then part of the aim of this chapter has been achieved. Neither the economic framework nor criminological theory presented is new. Putting the two explicitly together in this manner may be. In particular, the manner in which risk is constituted and influences the supply of criminal opportunities, and the roles of rewards and excuses, have significantly changed from economic models of crime that focused upon punishment. This is not to state that 'crime and punishment' models are incompatible with the present model. Rather, the present model places an emphasis upon different variables which are likely to suggest different policy implications. Although not formally mathematically proven in the way that Becker showed how punishment works, it is possible that formal mathematical development of the present analysis can be used to explore how prevention works. In time it could lead to analysis that derives optimal levels of prevention of different techniques and methods, and thus to recommendations for public and private policies against crime.

References

Becker, G. (1968) 'An Economic Approach to Crime and Punishment', *Journal of Political Economy*, 76, pp. 169–217.

Clarke, R. V. and Homel, R. (1997) 'A Revised Classification of Situational Crime Prevention Techniques', in S. P. Lab (ed.), *Crime Prevention at a Crossroads* (Cincinnati, OH: Anderson) pp. 35–57.

Clarke, R. V. (1997) 'Situational Crime Prevention', in R. V. Clarke (ed.), *Situational Crime Prevention: Successful Case Studies*, 2nd edn (Guilderland, NY: Harrow and Heston) pp. 1–18.

Cornish, D. and Clarke, R. V. (1985) 'Modelling Offenders' Decisions: a Framework for Research and Policy', in M. Tonry and N. Morris (eds), *Crime and Justice: An Annual Review of Research*, vol. 6 (Chicago, IL: University of Chicago Press), pp. 147–85.

—, — (1986) *The Reasoning Criminal* (New York: Springer-Verlag).

—, — (1987) 'Understanding Crime Displacement: an Application of Rational Choice Theory', *Criminology*, 25, pp. 933–47.

Field, S. and Hope, T. (1989) 'Economics and the Market in Crime Prevention', *Research Bulletin 26*, Home Office Research and Planning Unit, 40–4 (London: HMSO).

Pease, K. (1998) *Repeat Victimisation: Taking Stock*, Home Office Police Research Series Paper 90 (London: Home Office).

Sloan-Howitt, M. and Kelling, G. L. (1997) 'Subway Graffiti in New York: "Getting up" vs. "Meanin' it and Cleanin' it" ', in R. V. Clarke (ed.), *Situational Crime Prevention: Successful Case Studies*, 2nd edn (Harrow and Heston) pp. 242–9.

van Dijk, J. J. M. (1992) 'Understanding Crime Rates: On the Interactions Between Rational Choices of Victims and Offenders', *British Journal of Criminology*, 34, pp. 105–21.

Part II

Crime and the Labour Market:
Economic and Structural Factors

6

Editors' Introduction

In the preceding section we examined the economic approach to the aetiology of criminal offences and its implications for regulating criminal behaviour. We learned that the effort is organised around formal models whose key elements are consistent across different applications by different economists and that considerable effort has gone into testing the effects of these elements, so that, through iterative research, the model has been refined.

While the basic economic model has the merits associated with a formal, testable heuristic, we also know it is susceptible to problems of measurement and the difficulties of establishing comparable data across time and jurisdictions. These problems have harsher effects the more sophisticated the questions to which we present the model. It is logical that a model developed to understand the decision-making of individual potential offenders (but largely tested using macroeconomic data) should be applied to trend analysis of the relationship between broad economic indicators and crime rates. After all, the individual-level model understands crime as a form of work, with criminal and non-criminal forms of income-bearing behaviour being compared for their merits. Fluctuations in the state of the economy – notably those affecting the supply of legitimate employment and of goods – seem likely to impact on individual decisions which, when aggregated, give us our crime rates. A major application of economic analysis has therefore been in understanding the relationship between crime and the regional or national economy, particularly that between crime and the labour market.

In this section we examine studies which attempt to measure the impact of economic and structural factors on crime. Simon Field points out the long span of debate over the idea that crime has economic causes, and offers a useful introduction to the main relationships which have been examined. While debate has been most intense over the relationship between unemployment and crime, Field suggests that the strongest relationship is between per-capita real personal consumption and the crime rate (with interesting differences between personal and property crime). Field illustrates the long-term trend analysis used to study these relationships, and offers an account of why the unemployment-to-crime relationship does not show more strongly when researched empirically.

Crime and Consumption*

Simon Field

The idea that crime has economic causes has been debated since time immemorial. Quantitative studies on the issue stretch back to von Mayr's (1867) finding that there was more property crime in Bavaria at times when the price of rye was high, and lower at times when it was low.

Much of the discussion of the economic causes of property crime reflects two basic ideas. First, wealth may be an incentive to crime simply because with prosperity there are more goods available to steal. Second, wealth may be disincentive to crime because wealthy people have less need to steal. A large part of the theoretical literature on the relation between economic conditions and crime consists of an elaboration of these two basic ideas. For example, under the economic theory of crime developed by Becker and others, it is argued that potential criminals allocate their time to a mix of legitimate and illegitimate activities depending on the risks and rewards associated with each. Legitimate rewards will depend on wage rates. Illegitimate rewards will depend in part on the goods available to steal. Both wage rates and available goods will broadly depend on the state of the economy.

This double effect of wealth on crime makes theory difficult to test. It can be argued both that an increase in wealth should cause crime to rise, and that it should cause crime to fall. If theory is uncertain, then so have been the recent empirical findings, which have shown ambiguous evidence for a relation between unemployment and income levels of crime, although there is some more consistent evidence of a positive relation of inequality and crime (Box, 1987; Long and Witte, 1981; Orsagh and Witte, 1981).

A recent study of trends in recorded crime in England and Wales has cast some new light on the paradoxical role of wealth (Field, 1990). The main finding is clear, strong and capable of wide generalisation. The key economic factor found to be associated with rates of crime is 'per-capita real personal consumption' – the amount that each person in the country

*This chapter is reproduced by permission of the Controller of Her Majesty's Stationery Office.

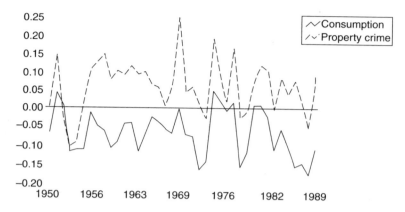

Figure 6.1 **Property crime and consumption: annual growth in recorded property crime and annual decline in personal consumption per capita (+ 3). Logarithmic growth rates**

spends, on average, in any year. With economic growth, personal consumption gradually increases, but its rate of growth varies according to the state of the economy. The research study found that during periods when personal consumption is growing relatively rapidly, property crime tends to grow relatively slowly or even fall. During periods when personal consumption is growing more slowly (or even falling) property crime seems to grow more rapidly.

Since consumption growth is inversely related to the growth of property crime, the negative of consumption growth – consumption decline – is positively related to increases in property crime. Since a positive relationship can be more vividly illustrated consumption decline is plotted in Fig. 6.1 against property crime growth, so that the parallel with crime growth can more readily be seen. The strength of the association between property crime and personal consumption is remarkable, particularly during the past 20 years. Thus, for example, the declines in property crime which took place in 1987–88 occurred in the context of historically high levels of consumption growth.

The finding is of wide applicability. It is separately demonstrable for residential and non-residential burglary, theft of and from vehicles, theft from shops, other theft, robbery and criminal damage. The association between property crime and personal consumption has held throughout the twentieth century in England and Wales, and a similar association can be demonstrated for the United States, Japan and France but not in

Sweden or West Germany. Economic factors in general, and consumption growth in particular, appear to be among the most important determinants of fluctuations in the growth of rates of property crime in industrialised countries.

The evidence suggests that this effect is *short term*, rather than one which might explain the growth of property crime in the long term. Thus, although property crime certainly tends to grow less quickly during years in which consumption grows rapidly, there is also evidence of compensating 'bounce-back' in the following years. Thus, the long-term rate of property crime growth might be unaffected by a surge in consumption growth. While the full relationship between economic growth and growth in property crime in the long term is as yet unclear, it seems that the effects identified in this study have only a limited bearing on the issue.

Personal crime – sexual offences and violence against the person (but not robbery) – also shows a distinctive relation to personal consumption, although the relation is weaker than is the case for property crime. Personal crime seems to grow more rapidly during periods when consumption is growing relatively rapidly. This means that personal crime responds to consumption growth in the *opposite* manner to that of property crime: during periods of slow consumption growth, personal crime tends to grow more slowly than usual, whereas in periods of rapid consumption growth, personal crime also tends to grow more rapidly.

This pattern, whereby economic boom periods are associated with small increases in property crime, but higher increases for personal crime, has been observed before. Bonger (1916), although he questioned the implied inevitability, refers to the frequent recognition that economic crimes (such as theft) occur in periods of economic depression, while crimes against the person – 'crimes of vengeance' – occur in periods of prosperity. In a statistical study of crime in England and Wales, Thomas (1925) found some inverse correlation over time between economic prosperity and property crime, and some indication of a positive correlation with offences against the person. Radzinowicz (1971) states that enquiries have shown that

> during periods of prosperity and depression, larceny and assaults move in converse directions. Crimes against the person (such as assault and battery) tend to go up during a period of prosperity and often decrease during a depression. A similar trend has been observed with respect to sexual offences.

Curiously enough, this idea has been somewhat eclipsed in more recent work on crime trends and the economy, perhaps because many of the more recent studies have tended to concentrate on unemployment. For example, Box (1987) surveyed the recent literature on recession and crime primarily in terms of measures of unemployment and income inequality, and their hypothesised (positive) effect on crime. Such studies are clearly important, but the evidence presented here suggests that an older theoretical tradition also has much to offer.

Explanations

Different factors could have positive effects on crime (such that an increase in the factor induces an increase in crime), or negative effects (such that an increase in the factor induces a decrease in crime). In principle, consumption growth might have three sorts of effect on crime by:

1. Increasing the goods available for theft or vandalism – the positive *opportunity* effect.
2. Providing an increasing capacity for the lawful acquisition of goods, thereby reducing the temptation of unlawful acquisition through theft – the negative *motivation* effect.
3. Altering the pattern of crime opportunities by precipitating an alteration in lifestyle or 'routine activities' – the positive *lifestyle* effect.

Property crime growth is negatively affected by consumption growth in the short run, suggesting that the motivation effect is dominant, so that rapid growth in consumption generates low growth in property crime. In the long run the opportunity and lifestyle effects appear to balance out the motivation effect, so that the long-run effect of economic growth on property crime may be small.

Why does only one effect – motivation – dominate in the short run? Potential victims are very widely spread in society. Potential offenders, however, are likely to be concentrated in particular social groups whose position in the labour market is liable to be weak or marginal. Fluctuations in aggregate consumption growth will therefore tend to be amplified in the experience of potential offenders. Better-placed labour market groups tend to be more insulated from the vicissitudes of the economy. It follows that a surge in consumption growth will be amplified in the experience of potential offenders, and the 'motivation' effect on them will therefore outweigh the 'opportunity' and 'lifestyle' effects on potential victims.

Property crime will therefore be lower than it otherwise would be. Conversely a deep trough in consumption growth will result in more property crime. In addition, a sharp fall in consumption growth may trigger frustration as economic expectations are lowered. This could undermine social controls. This theory provides an explanation of the observed relation of consumption and property crime, and explains why the effect is cyclical rather than long term.

Rates of personal crime – violence against the person and sexual offences – respond positively to rapid consumption growth. Personal crime is not directly affected by the goods available to the victim or the offender, but is affected by lifestyle and the pattern of routine activities, which in turn is affected by consumption and income growth. There is good evidence that people who go out more often are much more likely to be the victims of personal crime than those who do not, and there is also evidence that when consumption increases some of that consumption goes with increased time spent outside the home. This suggests an explanation for the observed positive relation between personal crime and consumption – when spending rises people are more often outside the home, and as a result there are more opportunities for personal crime.

Unemployment and Crime

Much of the research attention devoted to the relation of crime to the economy has concentrated on the potential effects of unemployment. Although a recent study suggested that young men are more likely to commit crimes when unemployed (Farrington *et al.*, 1986), aggregate studies – comparing the unemployment and crime rates among whole groups of people across place or time – have yielded very equivocal evidence of any relation between unemployment and crime.

The present study yielded compelling evidence that the rate of growth in property crime is closely related to the state of the economy, to which the unemployment rate is closely bound. Could the real aggregate relationship not therefore be with unemployment rather than with consumption? This seems unlikely. The relationship of consumption to property crime is an order of magnitude stronger than that of unemployment, and once the effect of consumption is taken into account, the unemployment rate adds nothing to the explanation of fluctuations in property crime growth. Although unemployment and consumption are both indicators of the state of the economy, consumption is a coincident indicator, whereas unemployment is a lagging indicator. Property crime, like consumption, appears to be a coincident indicator of the state of the economy. This means that in practice, as the economy – say – starts to emerge from a low point in the economic cycle, personal consumption will start to grow faster, and property crime to grow more slowly *before* unemployment starts to fall. Since causes must precede effects, the turning points in the crime figures cannot be the result of turning points in the unemployment figures.

This does not prove that unemployment cannot cause property crime, for the findings concern aggregate data. A small proportion of the population – essentially a subgroup of young males – are responsible for a large proportion of recorded crime. Unemployment trends among this group may differ markedly from those for the whole population, and could suggest a very different picture.

Conclusions

The implications of this study have both direct practical application and theoretical value. On the practical side the findings open up a novel way of looking at statistics on recorded crime. They suggest that national trends in recorded crime should be interpreted in the light of the prevailing economic circumstances, particularly personal consumption. Given that economic trends, including consumption, are commonly forecast two or three years into the future, trends in recorded crime may be forecast on that basis over a similar period. This means that the demands on the whole criminal justice system can be more effectively forecast. Moreover, the demands on later stages in the criminal process, such as court proceedings and prison receptions, are also linked to recorded crime and the economic factors which underlie it, although they also reflect many other factors. This means that economically based forecasts of further stages in the criminal process are also possible. This represents a valuable contribution to resource planning.

On the theoretical side, the finding demonstrates clearly that economic factors can have an important bearing on crime. While the nature of these factors requires further analysis, it suggests that attempts to reduce the level of crime could usefully address the economic circumstances of those perceived to be at risk of criminal involvement.

References

Bonger, W. (1916) *Criminality and Economic Conditions* (Chicago, IL: Little & Brown).
Box, S. (1987) *Recession, Crime and Punishment* (London: Macmillan).

Farrington, D. P., Gallagher, B., Morley, L., St Ledger, R. J. and West, D. J. (1986) 'Unemployment, School Leaving, and Crime', *British Journal of Criminology*, 26, pp. 335–56.

Field, S. (1990) *Trends in Crime and their Interpretation: A Study of Recorded Crime in Post-war England and Wales*. Home Office Research Study no. 119 (London: HMSO).

Long, S. K. and Witte, A. D. (1981) 'Current Economic Trends: Implications for Crime and Criminal Justice', in K. N. Wright (ed.), *Crime and Criminal Justice in a Declining Economy* (Cambridge, MA: Oelgeschlager, Gunn & Hain), pp. 69–143.

Orsagh, T. and Witte, A. D. (1981) 'Economic Status and Crime: Implications for Offender Rehabilitation', *Journal of Criminal Law and Criminology*, 72, pp. 1055–71.

Radzinowicz, L. (1971) 'Economic Pressures', in L. Radzinowicz and M. Wolfgang (eds), *Crime and Justice*, vol. 1 (London: Basic Books).

Thomas, D. A. (1925) *Social Aspects of the Business Cycle* (London: Routledge).

von Mayr, G. (1867), 'Statistik der gerichtlichen Polizei im-Konigreiche Bayern', cited in H. Mannheim (1965), *Comparative Criminology*, vol. 2 (London: Routledge).

7

Editors' Introduction

Field's introduction to the structural and longitudinal modes of analysis suggests the importance of employment, unemployment and goods as economic factors accounting for crime, along with the need for system-level data over long time periods (and alignment of measures). Richard Freeman elaborates these themes, taking as central the fact that, pace *Becker-type analysis, the US saw high rates of crime during the 1980s despite increased punitive sentencing. Freeman examines whether increased economic incentives to commit crime accounted for this, considering such factors as greater earnings inequality and the fall of real earnings experienced by the lower skilled worker, and the growth of illegal drug trafficking. As well as labour market incentives, Freeman examines the reverse effect of criminal activity on labour market outcomes, and the financial returns to crime (a surprisingly under-researched topic). He notes that, as the proportion of men with criminal records has increased, attention has turned from the effect of the labour market on crime to the reverse relationship, that of crime on labour market outcomes — for example, the way crime affects current and future employment and earnings.*

The chapter shows the eclectic sources on which economic analysis may draw, including not only time-series studies but ethnographic research, cross-sectional area studies, comparative studies of economic decisions at individual level, and social experiments. Freeman offers clear conclusions from research on the link between the job market and crime, suggesting that improving the job market for those prone to commit crime would have a marked effect on crime rates.

Crime and the Labour Market*

Richard B. Freeman

The question that has traditionally motivated analyses of crime and the labour market has been the effect of unemployment on crime. Many people believe that joblessness is the key determinant of crime, and have sought to establish a significant crime–unemployment trade-off. Studies through the mid-1980s found that higher unemployment was associated with greater occurrence of crime, though the unemployment–crime link was statistically looser than the link between measures of deterrence (such as the severity of criminal sentences or chances of being caught) and crime, and was more closely aligned to property crimes than to violent crimes.[1] Most important , although the rate of unemployment drifted upward from the 1950s to the 1990s, even the largest estimated effects of unemployment on crime suggest that it contributed little to the rising trend in crime.

Developments in the 1980s–1990s raise a broader set of issues regarding the link between the job market and crime.[2] The high rate of crime in the 1980s despite increased incarceration directs attention at potential increases in economic incentives to commit crime. Perhaps the widely heralded increase in earnings inequality and the fall in the real earnings of the less skilled men who commit most crimes gave young men a job market 'push' into crime. Perhaps the growth of the illegal drug business raised the returns to crime compared with those from work. At the minimum, the massive incarceration of criminals in the 1980s has brought the issue of crime from the periphery to the centre of discussions of poverty and the underclass.

In this chapter I examine evidence and studies regarding the effect of labour market incentives on crime, the reverse effect of criminal activity on labour market outcomes, and the financial payoff to crime. There are two 'bottom-line' questions: (1) What part, if any, of the high rate of criminal activity among young men results from the deteriorating job market for less skilled workers? And (2) How does crime affect the long-run economic position of those who commit crimes? Before turning to these questions I review the basic facts on the criminal participation of

*This chapter was originally published in Wilson and Petersilia (eds), *Crime* (1995), and is reproduced by kind permission of ICS Press.

young men and incarceration that make crime important to understanding the economics of the American 'underclass'.

The Facts

In 1993 the number of men incarcerated in the United States was 1.9 per cent of the number in the labour force. The number of men on probation or parole relative to the male labour force was approximately 4.7 per cent;[3] so that the number of men 'under supervision of the criminal justice system' was 6.6 per cent of the male workforce – one man incarcerated, probated, or parolled for every 12 men in the workforce. This was nearly as many men as were unemployed in that year. At extant growth rates the number under supervision will exceed the number unemployed in 1994–95.

No, I have not made an error. These figures do not refer to young men or to minority men. They refer to all men. For men aged 18–34, the ratio of those incarcerated to the labour force was 3.1 per cent; the number under supervision of the criminal justice system was 11 per cent of the workforce in 1993. For all black men the ratios to the workforce are 8.8 per cent incarcerated and 25.3 per cent under supervision relative to the workforce. For black men aged 18–34 the ratios to the workforce are 12.7 per cent incarcerated and 36.7 per cent under supervision.[4] Since a disproportionate number of prisoners are high-school dropouts, the proportion of less educated men, especially young men, who were incarcerated, probated, or parolled, was even greater (Freeman, 1992).

High though these figures are, they understate the extent to which American men are involved in criminal activity. Not everyone who commits crimes is caught by the police, and not everyone caught is convicted of crime. The magnitude of involvement in crime is such that analysts who once dismissed criminal behaviour as a peripheral issue to employment or poverty can do so no longer. No other advanced society has as large a proportion of its potentially productive workforce involved in illegitimate activities, nor as large a proportion incarcerated.

Trends

The most striking trend in crime statistics in the 1980s was the growth of the prison and jail population. From 1980 to 1991 the number of persons incarcerated rose at an exponential rate, with no sign of deceleration (Fig. 7.1). The average annual increase in the jail and prison population was 8.5 per cent. Had nothing else changed, the imprisonment trend should have greatly reduced the crime rate. It removed men with a high propensity to commit crime from society and increased the risk to

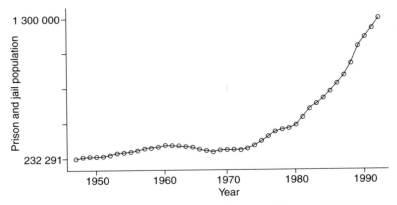

Figure 7.1   Prison and jail populations in the United States, 1947–92
Sources: Bureau of Justice Statistics, *Sourcebook of Criminal Justice Statistics 1991*;
Gilliard and Beck, *Prisoners in 1993*, Bureau of Justice Statistics Bulletin, NCJ-
147036. Estimates of jail population before 1983 based on prison population.

potential criminals that they would end up in jail or prison.

The standard administrative measure of crime, the Justice Department's
Crime Index, obtained from police departments around the country, does
not show the expected drop in crime. The Uniform Crime Rate, reflected
in the FBI's *Uniform Crime Reports* (UCR) and defined as the number of

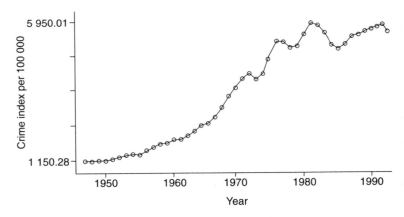

Figure 7.2   Uniform crime reporting index per 100 000, 1947–92
Sources: FBI, *Crime in the United States* (various years); Bureau of Justice Statistics,
Sourcebook of Criminal Justice Statistics 1991. Current statistics from the US
Department of Justice, Bureau of Justice Statistics.

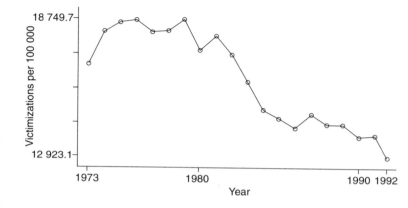

Figure 7.3 **Victimisations per 100 000, 1973–92**
Source: Bureau of Justice Statistics, *Criminal Victimization 1992*.

Uniform Crime Reporting Index crimes per 100 000 persons,[5] at best stabilised in the 1980s (Fig. 7.2). It fell from 1980 to 1984, then increased through 1991 to approach its 1980 peak level before dropping modestly in 1992. By contrast, the rate of criminal victimisations, defined as the number of times people report they or their family were victims of crime on the annual National Crime Victimisation Survey, dropped over the same period (Fig. 7.3), creating a problem of data inconsistency.

The victimisation figures differ markedly from the Uniform Crime Rate in level as well as trend.[6] Because individuals do not report all crimes to the police, reported victimisations range from 2.4 to 4.1 times the police data on crimes. In 1973, 32 per cent of victimisations were reported to the police. In 1991, 38 per cent of victimisations were reported to the police. A large proportion of the difference in volume of crime between the administrative data and victimisation survey is for crimes that are difficult to measure or report, such as rapes or larceny.

Several factors explain the difference in trends between the two sets of data. Some of the trend in the Uniform Crime Rate is due to an increase in the proportion of crimes that individuals report to the police. Boggess and Bound (1993) estimate that increased reporting accounts for about one quarter of the difference in trend. Perhaps another quarter of the difference in trend is the increase in victimless drug crimes, which individuals do not report. This still leaves a sizeable difference in trend. Should one put greater weight on the administrative UCR data or on the survey data on victimisation in assessing the trend in crime? One way to judge which data might be more accurate is to examine changes in crimes that are well measured, such as murder or automobile thefts. Murder rates roughly stabilised in the 1980s, rising for teenagers while falling for adults. Automobile thefts rose in the period. The change in these crimes suggests that the stability in the UCR rate may give a better fix on what is happening to crime levels than the falling rate in the victimisation survey.

Increased Propensity for Crime

As noted earlier, the rough doubling in the prison and jail population in the 1980s should, all else the same, have greatly reduced crime because of the incapacitation of criminals. It produced, in addition, an upward trend in the proportion of crimes that resulted in prison sentences (following a decline from the mid-1960s to the late 1970s) (Langan, 1991) that should have further reduced crime through the deterrent effect. The different trends in the UCR and victimisation rates notwithstanding, the 1980s levels of both statistics differ so much from the levels that massive incarceration should have produced that they tell the same story about criminal behaviour: namely that the *propensity for crime* among non-institutionalised men increased immensely in the 1980s.

Figure 7.4 demonstrates the increased propensity for crime in the UCR data. It plots the annual relation between the proportion of the adult male population confined to prison or jail and index crimes per man in the non-institutional population.[7] If the propensity for crime in the non-institutional male population were constant, the increased confinement would reduce crime through incapacitation or deterrence, producing downward-sloping confinement–crime (CC) curves. The greater the rate of criminal activity of those sent to jail or prison and the greater the deterrent effect of jail or prison on future crimes, the more steeply sloped will be the CC curve.

The curve joining the percentage confined and crimes per man in the figure is not, however, downward-sloping. It is a straight line, because the increased confinement of the population in the 1980s was accompanied by a roughly constant number of crimes per adult male. The three hypothetical CC curves in the figure show what 'should' have happened to the rate of crimes per adult male as a result of increased incapacitation of criminals. These curves take 1978 as a starting year and calculate hypothetical crime rates by subtracting from the number of crimes in each succeeding year different estimates of the change in crime resulting from the growth of the prison and jail population since 1978. The changes in crime are obtained from conservative estimates of the number of crimes each additional

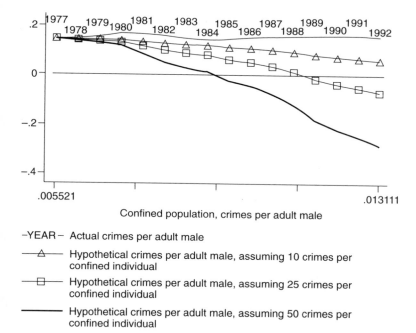

Figure 7.4   Crimes and confined population per adult male, 1977–92
Sources: FBI, *Crime in the United States* (various years); Bureau of Justice Statistics, *Sourcebook of Criminal Justice Statistics 1991*. Current statistics from Gilliard and Beck, *Prisoners in 1993*, Bureau of Justice Statistics Bulletin. Estimates of jail population before 1983 based on prison population. Population figures from *Economic Report of the President* (1993).

confinee would have committed had he been free,[8] and ignore deterrent effects that should have begun operating in the mid to late 1980s and the changing age structure of the male population,[9] both of which would further add to the expected drop in crime. By construction, the hypothetical CC curves slope downward. The gap between the actual and the hypothetical CC curves measures the increased propensity for crime among the non-institutional male population from the base 1978 year.

Figure 7.5 shows actual CC curves for reported victimisations committed per adult male and hypothetical curves calculated in a similar manner to those in Figure 7.4. The actual CC has a negative slope, reflecting the drop in victimisations. In calculating hypothetical CCs in this figure, I assume a greater number of crimes per person confined than I

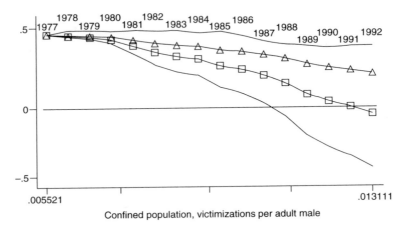

Figure 7.5   **Victimisations and confined population per adult male, 1977–92**
Sources: Bureau of Justice Statistics, *Criminal Victimisation 1992* and *Sourcebook of Criminal Justice Statistics 1991*; Gilliard and Beck, *Prisoners in 1993*, Bureau of Justice Statistics Bulletin. Estimates of jail population before 1983 based on prison population. Population figures from *Economic Report of the President* (1993).

did in Figure 7.4 because the volume of victimisations exceeds the volume of index crimes, though my estimates are still moderate ones. The hypothetical CC curves show a much more pronounced negative slope than the actual CC curve. The gap between the curves shows an increased propensity for crime comparable to that in Figure 7.4.

The bottom line is that *the propensity of the non-institutional population to commit crime rose sharply in the 1980s.*

How might we explain this increase? The economist is naturally drawn to a job market explanation. Given the well-documented growth of earnings inequality and fall in the job opportunities for less skilled young men in this period, and the increased criminal opportunities due to the growth of demand for drugs, the economist finds appealing the notion that the increased propensity for crime is a rational response to increased job market incentives to commit crime. What is appealing, however, need not be true or, if true, may be difficult to prove. To see how much weight we might reasonably give to an earnings explanation of the rising crime propensity, I turn next to extant studies of the effect of economic incentives on crime.

Labour Market Incentives and Crime: Statistical Studies

Social science analyses of the effect of the labour market on crime take several forms: time-series studies that compare the crime rate with labour market variables over time; cross-area studies that compare crime and economic characteristics across cities or states; and individual studies that compare crime and economic characteristics across people. In addition, there are longitudinal studies that follow the same area or individual over time, as economic opportunities change, and studies based on social experiments, in which the experimenter manipulates opportunities.

Studies of crime and the job market through the mid-1980s, which focused largely on unemployment, have been reviewed and summarised in detail in Freeman (1983) and Chiricos (1987). Building on those reviews for the earlier period, I concentrate here on ensuing work, and the 'trend' in research results. Rather than updating the score card of findings, I direct attention to specific studies that are either particularly innovative or particularly convincing.

Time-series Studies

Time-series data allow us to examine the effect of the business cycle on crime and to answer the question of what might happen to crime levels if overall job prospects improved or worsened on a short-term basis. For this reason, analysts often use time-series data to examine the effect of unemployment on crime. But time-series analyses suffer from a myriad of problems that make many social scientists wary of their results. Variables tend to move together over time, providing little independent variation from which to infer relations, and often suffer from a tendency for the unexplained part of the dependent variable to be correlated from one year to the next. All too often, addition of further observations or of another explanatory variable or choice of statistical technique, substantively changes results.

Time-series studies through the mid-1980s showed that the overall crime rate and the rates of particular crimes, such as burglary, were positively related to unemployment. But the estimated effect of unemployment was moderate and, as noted, incapable of explaining much of the upward trend in crime. Figure 7.6 shows a modest positive relation between the number of index crimes per adult male and

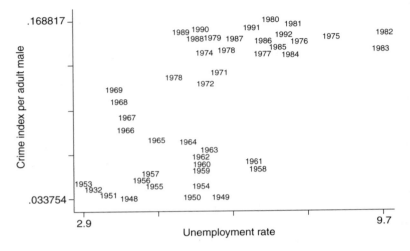

Figure 7.6   Crimes per adult male and unemployment, 1948–92
Sources: FBI, *Crime in the United States* (various years); Bureau of Justice Statistics, *Sourcebook of Criminal Justice Statistics 1991*. Current statistics from the Bureau of Justice Statistics. Population and unemployment figures from *Economic Report of the President* (1993).

unemployment in each year from 1948 to 1992, dominated by the upward trend in crime, so that any given unemployment rate is associated with very different crime rates over time. A linear regression of the crime rate per 100 000 in the population on a trend and the rate of unemployment gives a positive coefficient on the unemployment rate with a moderate standard error (that is, crime rate and unemployment rate are correlated). But the same regression with the crime rate per 100 000 adult men as in the figure (rather than per the entire population) gives a statistically insignificant positive coefficient on the unemployment rate.

Higher-tech statistical models – in which the *change* in a crime rate is regressed on the *change* in unemployment and the *change* in unemployment one year earlier – tell a more complex story about the relation between crime and unemployment. Calculations for the United Kingdom show that changes in the unemployment rate are associated with changes in crime in the same direction, consistent with the notion that unemployment raises crime, while changes in unemployment a year earlier have essentially no effect on crime (Hale and Sabbagh, 1991). But calculations for the United States show that changes in unemployment this year are associated with changes in crime in the opposite direction (an increase in unemployment reduces crime!), while the past year's change in the unemployment rate has the more plausible effect of changing crime in the same direction (Land, Cantor and Russell, 1994). Analysts have interpreted the negative relation between changes in current unemployment and crime as reflecting a reduction in criminal opportunities in a sluggish economy (when unemployment is high, there may be less to burgle and more people home watching their property) while interpreting the positive effect between changes in last year's unemployment and crime as reflecting the increase in criminal motivation due to joblessness. The net of the two effects varies by crime, is close to zero in several calculations, but shows that higher unemployment is associated with reductions in motor vehicle thefts.

The time-series results are, however, sensitive to the model and time period covered. Using a model with several explanatory variables over the period 1933–85, Cappel and Sykes report a positive effect of contemporaneous unemployment rates on crime rates for the United States. As a check on the robustness of the time-series relations I regressed changes in burglary rates on changes in unemployment rates and on changes in unemployment rates in the previous year, for the years from 1948 to 1993. I obtained a negative coefficient on the contemporaneous change in unemployment and a positive coefficient on the previous year's change, mimicking Land, Cantor and Russell (1994). However, the coefficient on the contemporaneous change in unemployment was insignificant, while the coefficient on the lagged change in unemployment was large and significant, implying that higher rates of unemployment are positively associated with crime.

All in all, I would not weigh heavily the time-series evidence. The same problems that plague time-series analyses of wages, interest rates and unemployment plague time-series analyses of crime. Differences in years covered or in the model chosen, or in the particular measures used affect results substantively. The safest conclusion is that the time-series are not a robust way to determine the job market – crime link. For more reliable results of how economic incentives may affect crime, I turn to evidence across areas and individuals.

Cross-section Area Studies

Studies of crime and the job market that use cross-section area data compare crime rates in areas with greater or lesser jobless problems or where the earnings of crime-prone groups or income inequality are particularly low or high. These studies are free from collinearity or serial correlation; but they suffer from their own set of inference problems. Areas may differ in labour market conditions and crime for reasons having to do with the features of the population that are not measured, producing spurious correlations or hiding true ones. In some 1960s cross-section studies, for instance, crime was inversely related to the percentage of non-whites in the area. At face value this would imply that non-whites are less likely to be criminals than whites, or that areas of black concentration are subject to less crime than areas of white concentration – both of which fly in the face of individual data on who commits crime, on who are the victims of crime, and on the locus of crime among neighbourhoods in a city. Rapid changes in the characteristics of areas, for instance a sudden boom or bust or change in demographic mix, may also give misleading inferences if crime (or other dependent variables) changes more gradually.

Still, cross-area studies are a natural way to examine the effects on crime of economic variables, such as income inequality or rates of poverty, that are likely to characterise the area for extended periods. At the minimum these data can answer such questions as: Is crime higher in areas with higher levels of income inequality or in areas with higher rates of poverty?

The majority of cross-area studies show a link between labour market factors and crime. In my 1983 review I classified four of 15 cross-area studies as giving significant effects for unemployment and an additional seven as giving positive but 'weak' results. Summarising 42 studies, including several for Canada or the United Kingdom, Chiricos reports coefficients on unemployment that were positive but insignificant in 51 per cent of the cases and positive and significant in 14 per cent of cases in pre-1970s data, and that were positive but insignificant in 44 per cent and positive and significant in 48 per cent of the cases in 1970s data (Chiricos, 1987: table 3, results for all crimes). Some of the cited studies use similar data (though processing it with different models), so that the results are not truly 'independent'. Some studies have larger samples and more precise estimates than others, so that simple counts of signs and significance of coefficients are also not ideal. Still, even without a definite mega-statistical analysis of these results, it is clear that the cross-area data support a positive unemployment – crime link.

Not all the work since those reviews has yielded statistically significant coefficients on unemployment, but nothing has arisen to overturn their conclusion.[10] As an exemplar study that extends the analysis to the 1980s, consider Lee's (1993) study of crime in 58 standard metropolitan statistical areas (SMSAs) from 1976 to 1989. He estimated the effect of economic variables on a set of crime rates using three statistical models: a cross-section model that compares economic incentives and crime among cities; a fixed-effects model where city dummy variables eliminate unmeasured city effects; and a model that allows for last period's unemployment to affect this year's crime rate. All three models gave a positive crime–unemployment link. In the cross-section analysis the overall crime and most specific crimes were positively associated with unemployment. In the fixed-effects model the total crime rate, property crime rate, burglary and motor vehicle theft rates were positively related to unemployment, while murder, rape and some other crime rates were not positively related to unemployment. The models which explored different time patterns of unemployment–crime effects confirmed the positive link between the variables. The magnitude of the link is, however, modest: Lee (1993) estimates that a one-point increase in the unemployment rate raises property crimes by 1.1 per cent to 1.4 per cent. This contrasts with a coefficient of variation in property crimes across SMSAs of roughly 30 per cent.

Results with respect to other labour market variables are also supportive of the notion that economic incentives affect crime rates. Some studies use the income of the population in an area and the percentage of families in poverty to measure the potential gain and opportunity cost of crime. Others include a Gini coefficient or other measure of inequality to capture both the gain and opportunity costs in a single term. The reviews by Freeman (1983) or Chiricos (1987) show that variables measuring inequality/poverty across areas are associated with differing crime rates across cities. Land, McCall and Cohen (1990) find that even homicide rates tend to be higher in cities with greater inequality. In his analysis of 127 SMSAs in 1979 and 1969, Lee obtained a significant positive relation between crime and inequality measured as the difference between the household income of the ninth decile and the first decile divided by the median household income, calculated from the Census of Populations for 1970 and 1980 (Lee, 1993). His model included numerous other controls, such as the percentage of an SMSA that was black, population density, and region of the country. Figure 7.7 gives the scatter plot between property crimes and inequality that underlies his work for 1979.

To what extent, if at all, can these cross-section findings explain the rising crime participation among adult men? From 1979 to 1990 the ratio of the difference between the ninth decile and tenth decile of household incomes divided by the median in the United States rose by about 12 percentage points.[11] Given the magnitude of Lee's estimated relation between crime and inequality, this change would have induced a 10 per cent increase in the crime rate. This goes part of the way to explaining why the Uniform Crime Reporting Index did not fall, despite rising incarcerations.

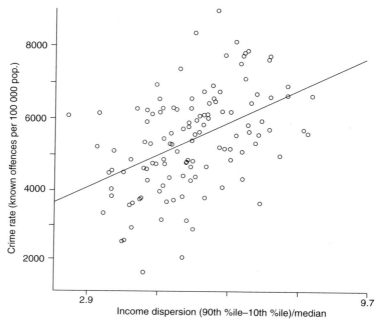

Figure 7.7   Property crime versus income inequality in 127 metropolitan areas in the United States, 1979
Source: David Sang-Yoon Lee, 'An Empirical Investigation of the Economic Incentives for Criminal Behavior' (1993).

But two aspects of Lee's analysis raise doubts about this inference. First, Lee finds that most of the inequality effect operated through a link between crime and income at the ninth decile: crime was more responsive to the income of the upper parts of the distribution than to income in the lower part of the distribution. This is troublesome because the rise of inequality took the form largely of falling real income in the lower part of the distribution. Second, he reports that changes in crime rates across the SMSAs from 1969 to 1979 were unrelated to changes in income inequality. Perhaps the cross-section pattern shown in Figure 7.7 is due to omitted area characteristics, and thus disappears when the analysis treats changes in variables. Alternatively, perhaps measures of changes in inequality among cities are subject to such huge measurement error that we should discount the results that show change over time.

In sum, cross-section evidence continues to support a positive link between unemployment and crime, and suggests that inequality may be an important contributor to crime. But there is enough statistical frailty in extant estimates to leave a door open to doubt about how helpful the cross-section inequality results will be in explaining the rising propensity to crime that characterised the 1980s.

Individual Comparisons

Studies that compare the economic circumstances of individuals who commit crimes with those who do not commit crimes, or the criminal behaviour of the same person in different economic circumstances, potentially offer the best way to assess how the job market affects crime. The main reason for this is that these studies focus on the people who are in fact making the crime decision, and on their particular circumstances. Some studies use records on arrests or on prisoners. Arrest or prisoner data accurately measure the characteristics of arrestees or prisoners, but do not provide information on criminals who have not been caught, nor on the characteristics of non-criminals. Other studies use self-reported criminal activity on household surveys. Survey data in which people self-report crime cover all criminal activity, whether the crime was solved or not, and include people who did not commit crime. But people may incorrectly self-report crime: black youths, in particular, understate their criminal involvement (Hindelang, Hirschi and Weiss, 1981).

The strongest evidence that economic incentives are important in determining the crime rate comes from studies of individuals. At a descriptive level, these studies find that criminals are disproportionately from the groups whose incomes and employment opportunities have been low and falling: young, less educated men, often with low scores on the Armed Forces Qualifying Test (AFQT) or other standard tests. The evidence also shows that those who end up in jail or under arrest are more likely to be jobless or to have low incomes than other groups. Two studies of the Philadelphia birth cohort of 1945 (Wolfgang, Figlio and Sellin, 1972) found positive relations between unemployment and crime. Tauchen, Witte and Griesinger (1993) report that youths who were employed for a larger percentage of a year were less likely to be arrested than those employed for a smaller percentage of a year; Thornberry and Christenson (1984) report a substantial contemporaneous positive relation of unemployment and crime. In the 1980 National Bureau of Economic Research (NBER) survey of inner-city black youth, 30 per cent of those who committed a crime held a job at the time of the survey, compared with 46 per cent of those who had not committed a crime (Freeman, 1987).

While this evidence makes it clear that the population of criminals overlaps with the population at the bottom of the increasingly unequal income distribution in the United States, it does not establish a causal link from the labour market to crime. The data may, after all, simply reflect the fact that the criminal population consists of people who are unable to succeed in society because of 'personal characteristics'. That is, the cause of both the poor labour market record and criminal activity may be a third variable having to do with the specific attributes of the individuals. If this were the case, improved labour market conditions would have little or no effect on the criminal's life of crime, although we would always find poor work records among criminals. Moreover, though it is hard to argue that wages have fallen among low-skilled workers because they engage in more crime than in the past, criminals may have higher *joblessness* than non-criminals because they rejected jobs in favour of unemployment – a status that enabled them to engage more readily in crime.

There are three ways in which researchers use information on individuals to surmount these problems and make plausible inferences about causality. One is to look at the same person in different periods. Farrington *et al.* (1986) compared the timing of criminal activity among young men in the United Kingdom. The question is whether these men were more likely to commit a crime when they were unemployed or when they held a job. Consider, for example, someone who is unemployed for six months and employed for six months in a year and who commits four crimes. If the person commits all the crimes while unemployed, it is reasonable to conclude that unemployment is associated with crime, not with some unobserved personal characteristic of the individual. By contrast, if the person commits the crimes as frequently when employed as when unemployed, we would reject the notion that his unemployment caused the crime. Farrington *et al.* find that crime rates are higher during periods of unemployment than during periods of employment. This does not 'prove' that the unemployment caused the crimes, but points in that direction.

A second way to link crime to economic incentives with data on individuals is to examine the relation of the individuals' criminal behaviour to characteristics of the area in which they live. The rate of unemployment or level of income in that area is presumably independent of the characteristics of the individual, and thus a good indicator of outside labour market incentives that might induce illegal activity. Analysis of the link between criminal behaviour and characteristics of the county in which a youth resides in the National Longitudinal Survey of Youth (NLSY) shows no relation between crime and unemployment and a positive relation between crime and the income level in the county (Lee, 1993). Good, Pirog-Good and Sickles (1986) find an insignificant negative relation between the monthly area unemployment rate and crime in a sample of 300 youths enrolled in a Youth Service Center in Philadelphia. Trumbull finds a negative coefficient on area unemployment in a sample of 2200 ex-offenders from North Carolina. These findings conflict with the results from the area studies, which find a positive relation between area crime and area unemployment. No-one has explored the reason for this divergence in results.

The third way to use individual data to infer causal links between economic factors and crime is to estimate labour supply relations between criminal participation and actual or predicted wages and criminal wages or perceptions of the attractiveness of crime. This form of analysis is infrequent because most data sets do not contain information on criminal behaviour or perceptions of returns or risks. An exception is the 1980 National Bureau of Economic Research Inner City Youth Survey, that included a special crime module designed to allow researchers to probe the economic model of criminal behaviour (Freeman and Holzer, 1986). Viscusi used these data to estimate the effects of personal objective factors and of perceptions of the return to crime on participation. He found that youths who believed that they 'make more on the street than on a legitimate job' were far more likely to engage in crime than others, and that estimated difference in income from crime and from legitimate work also significantly affected crime behaviour (Viscusi, 1986). His study does not 'prove' that these economic factors motivated crime. Perhaps those who commit crime feel it necessary to justify such by reporting that crime was lucrative. Still, this evidence supports that interpretation.

In a similar vein, Grogger (1994b) estimated the effect of 'potential' wages on criminal participation in the NLSY. He obtained a significant negative relation that implied that, roughly, a 10 per cent decrease in the real wages of youths would increase their crime rate by nearly 10 per cent. Applying this estimate to the observed drop in the real wages of young men, he predicts a 23 per cent increase in crime due to falling wages from the mid-1970s through the period 1985–88, which he notes is roughly equal to the 18 per cent increase in the index arrest rate for young males over that period. Grogger's estimates are imbedded in a highly structured economic model and may very well be sensitive to alternative specifications, so I would not take them as 'truth', but rather would see them as another piece of imperfect individual-level evidence on the role of economic factors in crime behaviour.

The Effect of Labour Market Incentives on Crime: Ethnographic Studies

Ethnographic studies of crime provide qualitative information on the way individuals view the opportunities and constraints in their local community and their perceptions of the factors that underlie the choice of crime or work, or both. By viewing events through the eyes of participants, ethnographic research can bring the decision to engage in crime 'up close and personal' and help us interpret statistical evidence.

The findings of recent ethnographic studies on youth gangs and crime provide strong support for a job market interpretation of the decision of young men to engage in crime. Jeff Fagan, who directed a major multi-site ethnographic study of youth gangs, reports that 'Gangs in South Central Los Angeles, Chicago, and Detroit changed in recent years from ethnic enterprises organised around turf, ethnicity, conflict, and natural group processes to business organisations with monetary and material goals. Money became the driving force and organisational principle for these groups.... [This] almost ideological emphasis on money ... [by gangs is] a dramatic shift from the gangs of ten and twenty-five years ago' (Fagan, 1992: chap. 9; 23, 25–6).

Here I summarise some of the conclusions from specific ethnographic studies that back up Fagan's conclusion about the economic factors in gang and crime activity:

All ... agreed money was the primary focal point within their gangs. Virtually all criminal activities are oriented toward this end.
(Vigil and Yun, 1990, on Vietnamese gangs, 156)

A dominant ... is their intense desire to obtain money for themselves.
(Vigil, Yun and Long, 1992: 49)

The 'gang as a business', albeit its illegal status, is a fact of life for many inner-city youths throughout America today. More and more young men are turning to the gang to make a living.
(Padilla, 1992, on Puerto Rican gangs in Chicago)

Gangs moved beyond the scavenger stage ... [to become] corporate gangs.
(Taylor, 1990: 112, on black gangs in Detroit)

Kids ... are drawn to the underground economy because of the opportunities that exist there. They know the work is hard and dangerous; there is no such thing as a quick dollar.
(Williams, 1989: 132)

Making money is their main motive.

> (Chin, 1990: 137, on Chinese gangs in New York)

Those who had joined a gang most often gave as their reason the belief that it would provide them with an environment that would increase their chances of securing money.

> (Jankowski, 1991: 60)

The conclusion that gangs survive and grow because of the financial rewards they gain for their members suggests that the resurgence of gangs is due largely to their potential to cash in on new illicit opportunities.[12] Ethnographers also report that many crime-prone youths disdain the types of low-wage work available to them: 'Simply wanting to work may not be enough, it is the type of work and wage they are willing to accept' (Quicker, Galesi and Batani-Khalfani, 1992: 4, 19). This observation is consistent with the economists' reservation wage–job mismatch story of inner-city joblessness (Holzer, 1986).

Taken together, the statistical and ethnographic evidence present a consistent story, one that supports the notion that crime responds to economic incentives.

The Effect of Crime on Labour Market Outcomes

As the number of men incarcerated or involved in crime has risen, attention has shifted from the effect of the labour market on crime to the reverse link – the effect of crime on labour market outcomes. How does crime affect current employment and earnings? How does it affect future employment and earnings?

Data on work and crime activity by individuals in the same period provide a way to answer the first question. As summarised earlier, the general finding in studies of individuals is that crime and joblessness go together. It is easy, however, to exaggerate the strength of the relation. Many people commit crimes while employed. In the NLSY the difference in the employment rate of young men who admit to committing crimes (but were not arrested or convicted or sent to jail) and those who did not so admit is rather low. In 1979 and 1980, 59 per cent of those who were out of school and unemployed said they committed a crime, compared with 53 per cent of those out of school and employed.[13] Tabulating the data the other way, of those who admitted committing crimes, 72 per cent had a job, compared with 76 per cent of out-of-school young men who did not have a job. Many young men hold jobs for short spells, and may move back and forth between legitimate and illegitimate earnings activities, as well as making money from crime and legitimate work at the same time. The big difference in employment rates between those who commit crimes and those who do not is found for young men who later go to jail for their crimes.

To determine which way the relation between crime and employment actually runs, Thornberry and Christenson estimated structural path models in which they sought to identify both the crime → unemployment and the unemployment → crime links. They found evidence for both (Thornberry and Christenson, 1984). While this is a plausible finding, I am leery of reading much into it. Without knowledge of what in fact influenced the individual's decision, which the data do not provide, any division of the relation between the variables is likely to depend critically on the particular structural model used to make the estimates.

Interpreting the causal link between criminal activity in one year and future labour market outcomes is much easier. Since the criminal activity precedes the outcomes, it is difficult to argue that this reflects the influence of job market opportunities on crime. It is easy, by contrast, to interpret any relation as reflecting the effect of crime today on future employment or earnings performance. Many employers eschew hiring persons with a criminal record (Finn and Fontaine, 1985). Some jobs legally exclude ex-offenders (Dale, 1976). On the supply side, individuals engaged in crime today are likely to build up criminal skills at the expense of legitimate skills, so that today's crime will alter the relative rewards from legal and illegal activity in the future.

In any case, studies of the effect of criminal activity on future job market outcomes in longitudinal survey data give clear results. Persons whose criminal behaviour leads them into prison have markedly lower employment rates in the future than those who do not commit crimes or those whose offences are more modest (Freeman, 1992; Hagan and Palloni, 1988; Sampson and Laub, 1994; Ferguson, 1994). My estimates show that a prison record has a substantial quantitative adverse effect on future employment: in the NLSY a young man incarcerated in 1979 worked about 25 per cent less in the ensuing eight years than a young man who had not gone to prison. In 1987, for example, respondents averaged 44 weeks of work over the year, whereas those who had been incarcerated in 1979 averaged 32 weeks of work. Part of this is due to the greater likelihood that these youths were in jail during some of the ensuing years. Sampson and Laub estimate a model that indicates that the effect of jail on job stability underlies much of recidivism. In the 1980 NBER survey I found that the monthly employment record of an individual deteriorates relative to that of others after a spell of prison.

By contrast, other involvement with the criminal justice system has much less, if any, long-term deleterious effect on employment. Grogger (1994a) reports only short-term effects of arrests on future employment using data for California. In the NLSY and two other data sets I found that anything short of probation has no discernible effects on future employment of youths (Freeman, 1992). Sampson and Laub interpret the strong effect of incarceration but not of committing crimes *per se* on future employment as supporting a developmental model of criminal activity. They suggest that a labelling theory in which individual behaviour is affected by their social label may help explain this result (Sampson and Laub, 1994).

The effect of imprisonment on future employment can be decomposed into two separable effects. The first is recidivism: persons who are incarcerated have a high recidivist rate and thus will be absent from the job market in many future years. The Department of Justice's 1983 follow-up of prisoners released in 1983 found that 63 per cent were arrested within three years of their release and that 41 per cent were reincarcerated. In a longitudinal sample that follows Georgia prison releases for 17 years, Needels finds that 61 per cent were reincarcerated, and that the releasees averaged 15 per cent of ensuing years in prison and nearly 30 per cent of ensuing years under supervision of the criminal justice system. The second effect is that when they are out of prison, ex-offenders work much less than 'otherwise comparable' men (Needels 1993).

Earnings from Crime

The reader will undoubtedly have noticed one missing element in my review of the statistical studies of the effect of economic factors on crime. Conceptually, the crime decision depends on how the earnings from crime (adjusted for the various risks of crime) compare with the earnings from legal activity. But most studies report estimates of the effect of criminal behaviour on unemployment, earnings inequality or predicted earnings rather than on the earnings from crime – *Hamlet* without the Prince, as it were.

The main reason is that few surveys ask about criminal earnings, and those that do may not obtain accurate estimates. Criminals are generally 'self-employed' – a group for whom it is difficult to obtain good data on earnings for *legal* activities. One must distinguish between gross and net earnings, and must determine the time the self-employed actually work to obtain an average wage rate comparable to wage rates in the job market. In contrast to workers paid a fixed wage per hour (or month) the self-employed are likely to have their hourly pay vary with the amount of time they work. Commit one burglary when you see a good chance and you do well per hour. Commit lots of burglaries and you are likely to move down the marginal returns curve, reducing average earnings.

To the extent that criminal earnings and legitimate earnings are positively correlated, the lack of a good measure of illegal earnings will bias *downward* the estimated effects of job market factors on criminal behaviour. The reason for this is that measures of legal opportunities will pick up both their posited negative effect on crime and the positive effect of correlated illegal opportunities.

Not surprisingly, given the data problem, there is disagreement on how much men make from crime and thus on the net payoff to criminal activity. In a 1992 study I reported estimates of earnings from crime from several surveys. The 1989 Boston Youth Survey showed sharply falling hourly earnings with numbers of crimes and relatively modest annual earnings of $3008 for 16–24-year-olds who report crime income. The 1980 NBER survey also showed modest annual earnings from crime (Viscusi, 1986). A 1990 survey of seven drug runners in Oakland estimated that hourly pay was $7.92. My conclusion from these scattered figures is that the hourly earnings exceed the hourly earnings the youths could make in legitimate work (which is consistent with the youths assessment, as well, on the Boston Youth Survey that asked if they can make more on the street or in legitimate work). At the same time, annual earnings from crime are modest, possibly because the marginal earnings fall sharply.

For adult criminals, my calculations for prisoners in the 1986 Justice Department Inmate Survey who said all of their income came from crime was that they earned $24 775 per year. Reuter, MacCoun and Murphy (1990) report that drug dealers earned $2000 a month net in their sample for Washington, DC, which they transformed into a $30.00 hourly rate of pay. Using the Rand Inmate Survey on numbers of crimes and various estimates of how much those crimes could garner, Wilson and Abrahamse estimated that criminal earnings for burglary/theft, robbery, swindling are modest, below the earnings these criminals could make at work, and find that only automobile theft is potentially profitable. They also report that criminals estimated their take from crime to be much higher than those crimes could plausibly have yielded, raising serious doubts about self-reported incomes from crime (Wilson and Abrahamse, 1992).

All told, the quality of data on criminal earnings is too weak for any strong claims about the long-term economic payoff to crime.

Conclusion

As the proportion of American men engaged in crime and incarcerated for criminal activity has risen, it has become increasingly important to understand the causes and consequences of criminal activity in addressing poverty as well as crime problems. While extant research leaves open many important questions, it has shown several important things about the link between the job market and crime:

1. In the 1980s and early 1990s the United States developed a large, relatively permanent group of young male offenders and ex-offenders, who for the most part are unlikely to be productive members of the workforce in the foreseeable future.
2. There is a general positive relation between joblessness and crime, which appears most strongly in comparisons of unemployment rates and crime rates across areas.
3. Labour market incentives beyond joblessness – the wages from legitimate work or measures of inequality – affect crime and potentially contributed to the rising propensity of non-institutionalised men to commit crime in the 1980s.
4. Incarceration reduces an individual's economic outcomes over the long run. This implies that the future costs of crime to an individual exceed the opportunity cost of devoting less time to legitimate activity today.

Although research has not yielded sufficiently strong results to predict reliably how much crime might fall if the job market for crime-prone groups improved substantively, the limited estimates we have are consistent with an expectation that such effects would not be negligible.

Notes

I have benefited from the research assistance of Ronald Chen.

1   See Freeman (1983) and Chiricos (1987).
2   Economics does not support the traditional focus on unemployment as *the* key labour market variable affecting crime. Rather, it posits that the decision to commit crime depends on the present value of economic returns to criminal activity compared with the present value of economic returns to legal activity. The returns to crime depend on: the chance of success, the money (utility) obtained from crime, less the value of the time spent at crime, the chance of being caught and convicted, the length of sentence and resultant earnings lost due to imprisonment. The crime decision should also depend on the effects of crime on future earnings opportunities and, because crime is risky, on attitudes towards the risks involved in crime, which range from risk of injury and death to risk of arrest, conviction and incarceration.
3   These figures are approximate because we do not have data for parole and probation for 1993, but must extrapolate 1990 figures.
4   These figures are larger than figures giving percentages of the various *populations* incarcerated or under supervision since not all adult men are in the workforce. I report figures relative to the workforce because my focus is on the links between crime and the labour market.
5   The Uniform Crime Reporting Index is based on statistics that local law enforcement agencies report to the FBI as part of the Uniform Crime Reporting Program. The crime index is based on seven crime categories: murder and non-negligent manslaughter, forcible rape, robbery, aggravated assault, burglary, larceny-theft and motor vehicle-theft, and arson.
6   Much of the information discussed here is taken from Boggess and Bound (1993).
7   I relate crimes to the male population, because the vast bulk of arrestees, prisoners and persons who self-report crime are men.
8   The numbers I use are much smaller than those in Zimring and Hawkins (1991: 95-6) or in Wilson and Abrahamse (1992: table 3).
9   In 1970 the proportion of the population that consisted of 15–34-year-old men was 14.7 per cent. In 1980 the proportion had risen to 17.6 per cent; but in 1990 it had fallen to 16.3 per cent.
10 The one contrary analysis that I have found is Trumbull's (1989) study of unemployment and crime across North Carolina counties, where he obtained a negative coefficient on the unemployment rate. But this does not mean that county data are inconsistent with more aggregate state or SMSA data: in an analysis of 120 counties in Kentucky, Howsen and Jarrell (1987) obtain positive coefficients on percentage unemployed or not in the workforce.
11 Calculated from United States Department of Commerce (1991: table B-2.)
12 While most ethnographies conclude that monetary incentives underlie gang activity, there is a general consensus that Chicano gangs are more turf-motivated (Moore, 1992; Jankowski, 1991; Vigil, 1990), and Jankowski also reports that Irish gangs in Boston are also more turf- than crime-business-oriented, in part because of connections with adults in the world of work that are missing in other communities.

13 The crime question is on the 1980 survey and refers to past crimes. We do not know the exact timing of the crime. I compare it with the employment status in 1979, but results are similar if I assume the crime was committed in 1980.

References

Beck, Allen J. and Shipley, Bernard E. (1989) *Recidivism of Prisoners Released in 1983*, Bureau of Justice Statistics Special Report, NCJ-116261 (Washington, DC: United States Department of Justice), April.

Boggess, Scott and Bound, John (1993) 'Did Criminal Activity Increase during the 1980s? Comparisons across Data Sources', National Bureau of Economic Research Working Paper 4431 (Cambridge, MA: National Bureau of Economic Research).

Bureau of Justice Statistics (BJS) (1992) *Sourcebook of Criminal Justice Statistics 1991* (Washington, DC: United States Department of Justice).

— (1993) *Criminal Victimisation 1992*, NCJ-144776 (Washington, DC: United States Department of Justice), October.

Cappel, Charles L. and Sykes, Gresham (1991) 'Prison Commitments, Crime and Unemployment: a Theoretical and Empirical Specification for the United States, 1933–1985', *Journal of Quantitative Criminology*, 7, pp. 155–99.

Chin, Ko-Lin (1990) 'Chinese Gangs and Extortion', in R. Huff (ed.), *Gangs in America* (New York: Sage).

Chiricos, Theodore G. (1987) 'Rates of Crime and Unemployment: an Analysis of Aggregate Research Evidence', *Social Problems*, 34, pp. 187–211.

Dale, Mitchell W. (1976) 'Barriers to the Rehabilitation of Ex-Offenders', *Crime and Delinquency*, 22, pp. 322–37.

Economic Report of the President (1993) (Washington, DC: United States Government Printing Office), February.

Elder, Glen H., Jr, Gimbel, Cynthia and Ivie, Rachel (1991) 'Turning Points in Life: the Case of Military Service and War', *Military Psychology*, 3, pp. 215–31.

Fagan, J. (1992) 'The Dynamics of Crime and Neighbourhood Change'. Presented at the Social Science Research Council Conference on the Urban Underclass, Ann Arbor, MI, 8–10 June.

Farrington, David P., Gallagher, Bernard, Morley, Lynda, St Ledger, Raymond J. and West, Donald J. (1986) 'Unemployment, School Leaving and Crime', *British Journal of Criminology*, 26, pp. 335–56.

Federal Bureau of Investigation (FBI) (various years) *Crime in the United States* (Washington, DC: United States Department of Justice).

Ferguson, Ronald F. (1994) 'Tables for Discussion of Young Black Males and Work at US Department of Labor'. 20 July.

Finn, R. H. and Fontaine, Patricia A. (1985) 'The Association between Selected Characteristics and Perceived Employability of Offenders', *Criminal Justice and Behavior*, 12, pp. 353–65.

Freeman, Richard B. (1983) 'Crime and Unemployment', in J. Q. Wilson (ed.), *Crime and Public Policy* (San Francisco, CA: ICS Press).

— (1987) 'The Relation of Criminal Activity to Black Youth Employment', *Review of Black Political Economy*, 16, pp. 99–107.

— (1992) 'Crime and the Employment of Disadvantaged Youth', in G. Peterson and W. Vroman (eds), *Urban Labor Markets and Job Opportunity* (Washington, DC: Urban Institute Press).

— and Holzer, Harry J. (eds) (1986) *The Black Youth Employment Crisis* (Chicago, IL: University of Chicago Press for National Bureau of Economic Research).

Gilliard, Darrell K. and Beck, Allen J. (1994) *Prisoners in 1993*, Bureau of Justice Statistics Bulletin, NCJ-147036 (Washington, DC: United States Department of Justice), June.

Good, David H., Pirog-Good, Maureen A. and Sickles, Robin C. (1986) 'An Analysis of Youth Crime and Unemployment Patterns', *Journal of Quantitative Criminology*, 2, pp. 219–36.

Grogger, Jeffrey (1994a) 'Criminal Opportunities, Youth Crime and Young Men's Labor Supply'. Department of Economics, University of California, Santa Barbara, February.

— (1994b) 'The Effect of Arrests on the Employment and Earnings of Young Men', *Quarterly Journal of Economics*, cx (1), pp. 51–72.

Hagan, John (1993) 'The Social Embeddedness of Crime and Unemployment', *Criminology*, 31, pp. 465–91.

— and Palloni, Alberto (1988) 'Crimes as Social Events in the Life Course: Reconceiving a Criminological Controversy', *Criminology*, 26, pp. 87–100.

Hale, Chris and Sabbagh, Dima (1991) 'Testing the Relationship between Unemployment and Crime: a Methodological Comment and Empirical Analysis Using Time Series Data from England and Wales', *Journal of Research in Crime and Delinquency*, 28, pp. 400–29.

Hindelang, Michael, Hirschi, Travis and Weiss, Joseph G. (1981) *Measuring Delinquency* (Beverly Hills, CA: Sage).

Holzer, Harry (1986) 'Black Youth Nonemployment: Duration and Job Search', in R. B. Freeman and H. Holzer (eds), *The Black Youth Employment Crisis* (Chicago, IL: University of Chicago Press for National Bureau of Economic Research).

Howsen, Roy M., and Jarrell, Stephen B. (1987) 'Some Determinants of Property Crime: Economic Factors Influence Criminal Behavior but Cannot Completely Explain the Syndrome', *American Journal of Economics and Sociology*, 46, pp. 445–57.

Jankowski, M. (1991) *Islands in the Street: Gangs and the American Urban Society* (Berkeley, CA: University of California Press).

Land, Kenneth C., Cantor, David and Russell, Stephen T. (1994) 'Unemployment and Crime Rate Fluctuations in the Post-World War II United States: Statistical Time Series Properties and Alternative Models', in J. Hagan and R. D. Peterson (eds), *Crime and Inequality* (Stanford, CA: Stanford University Press).

— McCall, Patricia L. and Cohen, Lawrence E. (1990) 'Structural Covariates of Homicide Rates: are There Any Invariances across Time and Social Space', *American Journal of Sociology*, 95, pp. 922–63.

Langan, Patrick (1991) 'America's Soaring Prison Population', *Science*, 251, pp. 1568–73.

Lee, David Sang-Yoon (1993) 'An Empirical Investigation of the Economic Incentives for Criminal Behavior'. BA thesis in economics, Harvard University, March.

Long, S. K. and Witte, A. (1981) 'Current Economic Trends: Implications for Crime and Criminal Justice', in K. N. Wright (ed.), *Crime and Criminal Justice in a Declining Economy* (Cambridge, MA: Oelgeschlager, Gunn & Hain).

Messner, Steven F. (1983) 'Regional and Racial Effects on the Urban Homicide Rate: the Subculture of Violence Revisited', *American Journal of Sociology*, 88, pp. 997–1007.

Moore, J. (1992) 'Institutionalised Youth Gangs: Why White Fence and El Hoyo Maravilla Change So Slowly'. Presented at the Social Science Research Council Conference on the Urban Underclass, Ann Arbor, MI, 8–10 June.

Needels, Karen (1993) 'Go Directly to Jail and Do Not Collect? A Long-Term Study of Recidivism and Employment Patterns Among Prison Releasees', Department of Economics, Princeton University, November.

Padilla, F. (1992) 'Getting into the Business', presented at the Social Science Research Council Conference on the Urban Underclass, Ann Arbor, MI, 8–10 June.

Quicker, J., Galesi, Y. and Batani-Khalfani, A. (1992) 'Bootstrap or Noose: Drugs in South Central Los Angeles'. Presented at the Social Science Research Council Conference on the Urban Underclass, Ann Arbor, MI, 8–10 June.

Reuter, Peter, MacCoun, Robert and Murphy, Patrick (1990) *Money from Crime* (Santa Monica, CA: Rand Drug Policy Research Center).

Sampson, Robert J. and Laub, John H. (1994) 'A Life-course Theory of Cumulative Disadvantage and the Stability of Delinquency', in T. P. Thornberry (ed.), *Developmental Theories of Crime and Delinquency: Advances in Criminological Theory*, vol. 6 (New Brunswick, NJ: Transaction).

Tauchen, Helen, Witte, Ann Dryden and Griesinger, Harriet (1993) 'Criminal Deterrence: Revisiting the Issue with a Birth Cohort', National Bureau of Economic Research Working Paper 4277, February.

Taylor, C. (1990) 'Gang Imperialism', in C. R. Huff (ed.), *Gangs in America* (New York: Sage).

Thornberry, Terence P., and Christenson, R. L. (1984) 'Unemployment and Criminal Involvement: an Investigation of Reciprocal Causal Structures', *American Sociological Review*, 49, pp. 398–411.

Trumbull, William N. (1989) 'Estimations of the Economic Model of Crime Using Aggregate and Individual Level Data', *Southern Economic Journal*, 94, pp. 423–39.

United States Department of Commerce (1991) *Money Income of Households, Families and Persons in the United States: 1990*, Current Population Reports, Series P-60, no. 174.

— (1994) Current statistics cited by telephone.

Vigil, J. (1990) 'Cholos and Gangs: Culture Change and Street Youth in Los Angeles', in C. R. Huff (ed.), *Gangs in America* (New York: Sage).

— and Yun, S. (1990) 'Vietnamese Youth Gangs in Southern California', in C. R. Huff (ed.), *Gangs in America* (New York: Sage).

— Yun, S. and Long, J. S. (1992) 'Youth Crime, Gangs and the Vietnamese in Orange County', presented at the Social Science Research Council Conference on the Urban Underclass, Ann Arbor, MI, 8–10 June.

Viscusi, W. Kip (1986) 'Market Incentives for Criminal Behavior', in R. B. Freeman and H. Holzer (eds), *The Black Youth Employment Crisis* (Chicago, IL: University of Chicago Press for National Bureau of Economic Research).

Williams, T. (1989) *The Cocaine Kids* (Reading, MA: Addison Wesley).

Wilson, James Q., and Abrahamse, Allan (1992) 'Does Crime Pay', *Justice Quarterly*, 9, pp. 359–77.

Wolfgang, Marvin E., Figlio, Robert M. and Sellin, Thorsten (1972) *Delinquency in a Birth Cohort* (Chicago, IL: University of Chicago Press).

Zimring, Franklin E. and Hawkins, Gordon (1991) *The Scale of Imprisonment* (Chicago, IL: University of Chicago Press).

8

Editors' Introduction

Among the methods used in the studies Freeman reviewed is the panel study. Its value is increasingly recognised across the social sciences, with important panel studies established on an ongoing basis for the study of trends. Witte and Tauchen offer a good instance of the method in research on the relationship between crime and work in a cohort study of young men in the USA. Their conclusions indicate the anti-criminogenic effects of pursuing conventional routes to economic security.

Witte and Tauchen find that working, and pursuing educational qualifications, not only decrease the probability of committing crime but do so in almost identical measure. But there appears to be an aptitude desideratum as well, because they find higher IQ and participation in parochial school education relates to lower criminal proclivities but no significant effect associated with award of a high school diploma. Moreover, the benign effects of employment and/or education are not reducible to the higher earnings potential associated with employment or educational attainment. This is consistent with other research, which provides scant evidence of significant wage or income effects on crime. The effects of employment and education on participation appear to be intrinsic in this respect. Witte and Tauchen consider that these findings suggest a need to revise static crime-as-work or demand-type models. This implies a move to more sophisticated multi-factor models where economic considerations register but do not operate in isolation, and to dynamic models which respond to maturation and career effects as well as background economic trends.

Work and Crime: an Exploration Using Panel Data*

Ann D. Witte and Helen Tauchen[†]

1 Introduction

In this chapter we consider the legitimate and criminal activities of young men. We are interested primarily in the relationship between young men's criminal activity and their employment. We use data for a birth cohort sample that contain information of the activities of a representative sample of young men over a seven-year period.

Beginning with Ehrlich's (1973) extension of Becker's (1968) pioneering work, economists have sought to model the nature of the relationship between employment and crime. These economic models generally see crime as like employment in that it takes time and produces income. The simplest of these models predicts that crime and work are substitutes. The implication is that increasing the availability of jobs and improving wages lowers the level of criminal activity. Empirical researchers have been unable, however, to provide strong, consistent and convincing evidence in support of this theoretical proposition.

*This chapter was originally published in Werner W. Pommerehne (ed.), *Public Finance and Irregular Activities*, Proceedings of the 49th Congress of the International Institute of Public Finance/Institut International de Finances Publiques (Berlin, 1993) supplement, vol. 49 (1994), pp. 155–67, and is reproduced by kind permission of *Public Finance/Finances Publiques*.

[†]The ordering of the authors' names was decided randomly; all work was done jointly. We would like to thank Marvin E. Wolfgang of the Center for Studies in Criminology and Criminal Law at the University of Pennsylvania for providing us with data, and Neil Weiner, also at the Center, for invaluable assistance with the data and for information about the Philadelphia criminal justice system. Terence Thornberry, Dean of the School of Criminal Justice, State University of New York, Albany, also provided helpful insights regarding the data and their use. Kevin Lang, Daniel Nagin and Robin Sickles provided valuable comments on earlier drafts. Finally, we would like to thank the National Science Foundation for the funding that made the research possible.

In light of these findings, we stepped back and surveyed empirical work, including ethnographies, on crime. This work provides important stylised facts. First and foremost, crime is a young man's game. In 1990, 70 per cent of the individuals arrested in the United States were between 16 and 34, and over 80 per cent of those arrested were male. Second, during their prime crime years (the late teens and early twenties), young men are often actively involved in educational pursuits. Only 45 per cent of the population between 16 and 19 years of age was employed in 1990. Third, ethnographic studies find that many criminal activities require relatively little time and are often combined with employment or education. In addition, participation in criminal activities has many distinctive attributes (e.g. flexible hours, immediate gratification rather than a weekly pay cheque, independence, flashy lifestyle) that makes it attractive to young men.

In this chapter we develop a model that reflects the above stylised facts, and we use a data set that has a number of unique elements. Our data contain information on the criminal, work and educational activities of a representative sample of young males in a large US urban area. The study traced the activities of the young men from ages 19 to 25. This is the period during which there is the greatest mixing of work, schooling and criminal activity.

Our work differs from most previous research in that we use data for a general population group, control for deterrent effects and consider educational as well as employment activities of young men. Most work on the relationship between employment and crime has used data for 'high-risk' populations such as prison releasees. The work either ignores deterrent effects or considers only specific (the effect of punishment on the individual punished) and not general (the effect of punishment on individuals without contact with the criminal justice system) deterrent effects. Much of the work considers employment but not educational activities.

The organisation of the remainder of the chapter is as follows. In the next section we review the empirical literature. In Section III we present the model that structures our empirical work and in the following section we describe the data and the empirical model. Section IV contains our empirical results and the final section our conclusions. To preview briefly our results, we find evidence that both employment and going to school are associated with less crime. The effects of schooling and employment on crime are virtually identical.

II   The Literature

Most theoretical models of crime are single-period individual choice models.[1] These models generally see the individual as deciding how to allocate time with criminal activity as one possible time use. Criminal activity is represented as being similar to employment in that it requires time and produces income. For convenience we refer to such models as 'crime as work' models.

The bulk of empirical work by economists has used aggregate data on crime rates, usually data obtained from the FBI's Uniform Crime Reports (UCRs), to estimate crime as work models.[2] This work has been criticised for aggregation bias, arbitrary identifying restrictions and poor data.[3]

Beginning in 1980 a small, but increasing, number of studies used individual, generally cross-sectional data, to estimate economic models of crime (e.g. Good, Pirog-Good and Sickles, 1986; Montmarquette and Nerlove, 1985; Myers, 1983; Viscusi, 1986a, 1986b; Schmidt and Witte, 1984; Witte, 1980). Since this work relates most closely to our own, we concentrate our review on this literature. For completeness we also discuss briefly some relevant work by sociologists (e.g. Rossi, Berk and Lenihan, 1980; Thornberry and Christenson, 1984) and psychologists (e.g. Farrington *et al.*, 1986; Gottfredson, 1985).

In the first half of the 1980s a number of economists used data for prison releasees to estimate models of criminal activity (e.g. Witte, 1980 and Myers, 1983). This work often had limited information on employment activities and no information on educational activities. For example, Witte uses the length of time required for releasees to find a job and the wage on the first job after release to reflect work activities. She has no information on educational activities. Further, studies using cross-sectional data for prison releasees cannot reveal how legitimate activities affect the decision to commence criminal activity. They can only reveal the effect of employment on the resumption of such activities.

Recent work using individual data has attempted to overcome some of the difficulties outlined above. Montmarquette and Nerlove (1985) and Thornberry and Christenson (1984) use data for general population groups. Farrington *et al.* (1986), Good, Pirog-Good and Sickles (1986), Gottfredson (1985) and Thornberry and Christenson (1984) use data that contain observations for at least two time periods. Only Farrington *et al.* (1986) and Thornberry and Christenson (1984) have panels that extend over a number of years (four years in each instance). These studies, however, do not use panel data estimation techniques. In addition, many of the studies fail to take account of the qualitative or limited nature of measures of criminal activity.

Furthermore, existing work with individual data does not control for possible deterrent effects arising from the actions of the criminal justice system. Indeed, some studies (e.g. Rossi, Berk and Lenihan, 1980;

Thornberry and Christenson, 1984) contain no deterrence variables.[4] Other studies address specific but not general deterrence issues by including variables that reflect the individual's perceptions or past experience with the criminal justice system (e.g. Montmarquette and Nerlove, 1985; Myers, 1983; Schmidt and Witte, 1984; Viscusi, 1986a, 1986b; Witte, 1980).

III Conceptualisation

As is usual, we assume a Von Neumann-Morgenstern decision maker. In contrast to existing work, we see the individual as choosing a level of criminal activity, denoted c, rather than the time to allocate to crime. We choose this approach because studies indicate that most criminal acts are unplanned[5] and that crime commission does not take much time (e.g. Hirschi, 1969; Crowley, 1981). We allow for the possibility of non-monetary gains from crime by entering the gains from crime, denoted $R(c)$, directly into the utility function. This reflects the fact that many offences yield no direct monetary gain (e.g. assault, drug use) and that even major property crimes produce surprisingly low monetary returns (Petersilia, Greenwood and Lavin, 1977; Swanson and Tabbush, 1988; Viscusi, 1986a).[6] Formally, the individual's utility with income from legal activities I, offence level c, and sanction s is

$$U[I, R(c), s; a^0] \qquad (1)$$

where a^0 denotes a vector of exogenous variables systematically related to preferences.

To reflect substantial evidence that criminal justice system actions depend on the number and type of offences,[7] we posit functions that relate the probability that an individual is arrested and the sanctions to the extent of the person's criminal activity. These functions may shift because of differences in individual abilities to avoid arrest and to mitigate punishment, denoted a^1, and because of exogenous changes in the criminal justice system (e.g. the availability of resources, administrative policy and the legal code), denoted b. Let $P(c; a^1, b)$ denote the probability of arrest and $S(c; a^1, b)$ the level of sanctions.

Our conceptualisation of criminal justice actions differs from that in most economic models of crime by seeing such actions as dependent on both the individual's criminal activity and exogenous shift factors. This has important implications for empirical work. For example, consider the implications of the model for how to measure the arrest probability in empirical work. As represented in the model above, an individual contemplating a criminal act does not face a single probability of arrest. Rather, the individual faces a schedule or function that relates each possible level of criminal activity to a probability of arrest. As pointed out by Cook (1979) and Poterba (1987), it is possible to estimate the effect of the probability schedule on crime only if there are exogenous shifts in the schedule. Under the model developed above, such exogenous shifts occur because of differences in individual abilities to avoid arrest and because of differences in police resources, police administration and the legal code. Variables reflecting such shifts and not the probability of arrest enter the crime equation.

For notational simplicity we ignore the possibility of multiple arrests and assume that in any time period the individual is either arrested once or not at all.[8] The individual chooses the level of criminal activity to maximise expected utility given by

$$EU = PU\,[I, R(c), S(c; a^1, b); a^0] + (1 - P)\,U\,[I, R(c), 0; a^0]. \tag{2}$$

With this model the optimal level of criminal activity, c^*, depends on total income from legal activities, the preferences of the individual, and exogenous factors that cause the probability or sanctions functions to shift. We do not view criminal activities and employment as being jointly determined. Crime and employment are not necessarily substitute time uses. This is consistent with findings that crime and work are often combined (e.g. Holzman, 1982; Phillips and Votey, 1984).

IV The Data and Empirical Model

Our primary data are for a 10 per cent random sample of males born in 1945 and residing in Philadelphia between their 10th and 18th birthdays. We combine this individual information with data on the total number of offences, police budgets, macroeconomic indicators and neighbourhoods in Philadelphia. Information was collected from school records, draft registration records, the Philadelphia Police Department, the FBI, a compendium on city government finances, the Philadelphia Community Renewal Program, and interviews carried out in 1970–1.

Using these data we created two panels, one a seven-year panel that traces cohort members' activities from 1964 through 1970 and the other an eight-year panel that ends in 1971. Since almost half of the interviews were conducted in 1970, the seven-year panel contains approximately twice the number of observations as the eight-year panel. The results discussed in the chapter are for the seven-year panel. Results for the eight-year panel and more details on the data are available in Tauchen, Witte and Griesinger (1988).

To estimate our model we require measures of the total income from legal activities, the preferences of the individual, and exogenous factors that cause the probability or sanction functions to shift. There are two primary measurement issues for this study, namely how to measure the level of criminal activity and how to reflect the criminal justice system actions. We use two measures of criminal activity. The first is a binary measure for whether or not the individual was arrested during the year. Although this is one of the most commonly used measures of criminal activity, it is well-recognised that such binary variables do not reflect the seriousness of the crimes committed or even the frequency of arrests. Given the differences in the types of crimes committed by our sample members,[9] we also used Sellin and Wolfgang's (1964) crime seriousness index to measure the level of criminal activity. Their index assigns a seriousness score to each crime for which the individual was arrested, and the seriousness scores for all arrests during the year are summed to obtain the crime index for the year.

The second measurement issue relates to criminal justice system actions and general deterrence. Both our model and empirical evidence indicate that criminal justice system actions depend on the level of criminal activity, on the individual's ability to avoid arrest and on exogenous factors related to the criminal justice system. Since the actions of the criminal justice system depend on an individual's criminal decisions, the individual's own experience with the criminal justice system cannot be used as a general deterrence variable. Most of the observed variation in individuals' experiences with the criminal justice system results from differences in crime seriousness and crime frequency, not from exogenous criminal justice system actions. Appropriate measures of general deterrence are numerical representations of exogenous changes in the criminal justice system. These variables must reflect variation that does not depend on the type and extent of criminal activity. Appropriate measures include changes in criminal justice resources and policies. Since there were no major changes in criminal justice system policies in Philadelphia during the study period, we use a number of variables related to the level of resources available (e.g. the real police budget per offence and per capita) as our measures of general deterrence.

We are not able to measure the income from legal activities directly since there are no income or wage variables in our data set. We have information, however, on the time allocated to work and on factors generally correlated with wages (i.e. IQ and a binary for whether or not the individual received a high-school degree) and incorporate these variables to reflect income from legal activities. Our data set also contains information on the time allocated to school during each year, and we incorporate this variable in order to control for educational activities.

The variables related to preferences are of three types: (1) variables that reflect family and community backgrounds (i.e. a binary equal to one if both parents were born in the US, a measure of the occupational status of the household head when the boy was in high school, a binary equal to one if the boy attended primarily parochial schools, the number of addresses during the school years, average income in the neighbourhood of residence during high school); (2) variables reflecting personal characteristics (i.e. IQ, a binary equal to one if the individual is white); (3) variables reflecting activities that occurred during the juvenile or young adult years (i.e. three variables indicating the type of charge at first arrest, the number of police contacts as a juvenile, the percentage of juvenile police contacts resulting in formal criminal justice system processing, a binary equal to one if the individual is married, and a binary equal to one if the individual was a member of a gang as a juvenile). Finally, we include a variable for the year of the panel to reflect the ageing of the cohort and other trend factors.

Most variables likely to reflect differing abilities to avoid arrest (e.g. intelligence or like-minded friends) are also likely to be related to differences in the individual's 'taste' for crime. To reflect this confounding of effects we interpret the coefficients on such variables as reflecting some mixture of preference and deterrence effects.

We estimate the models for the binary measure of criminal activity using a random effects probit model. The two-factor random effects probit model is an extension of the usual probit model. In the two-factor random effects models the disturbance terms are correlated across time for any individual but not across individuals. The component of the disturbance term that is correlated across time for any individual reflects unmeasured, persistent individual effects. If the error is uncorrelated with the explanatory variables, the parameter estimates of the error component probit model are consistent and asymptotically efficient (Chamberlain, 1984). See Tauchen, Witte and Griesinger (1994) for details.

V Empirical Results

Table 8.1 contains empirical results for the binary measure of criminal activity. The first column is for a specification including only variables that are unaffected by an individual's criminal or time allocation decisions

Table 8.1   Results for the probability of offending (asymptotic 't-ratios' in parentheses)

Independent variable	Model 1	Model 2	Model 3
General deterrence			
Real police resources per index offence	−0.0192***	−0.0197***	−0.0164***
	(−2.75)	(−2.79)	(−2.38)
Total Legal Income			
IQ	−0.0156**	−0.0163***	−0.008
	(−2.21)	(−2.00)	(−1.15)
Fraction of year individual was employed			−0.0075***
			(−3.40)
Fraction of year individual was in school			−0.0104***
			(−3.32)
Binary equal to 1 if received a high-school degree			−0.2131
			(−1.33)
Age/Returns to Legal Activity			
Year	−0.0028	−0.0009	0.0075
	(−0.09)	(−0.03)	(0.21)
Family background			
Binary equal to 1 if parents US born	0.3068	0.2735	0.1980
	(0.93)	(0.98)	(0.79)
Occupational status of household head during high school	−0.0048	−0.0065	−0.0034
	(−1.18)	(−1.54)	(−0.94)
Number of addresses during primary and secondary school	0.1116**	0.0138	0.0173
	(2.17)	(0.31)	(0.41)
Binary equal to 1 if attended parochial high school	−0.3823*	−0.4058*	−0.2656
	(−1.71)	(−1.79)	(−1.37)
Neighbourhood characteristics			
Average income in neighbourhood during high school ($1000)	0.0797	0.0002	0.0002
	(0.68)	(1.45)	(1.48)
Binary equal to 1 if high school neighbourhood predominantly Italian	0.0739	0.0070	0.0308
	(0.32)	(0.03)	(0.16)
Personal characteristics			
Binary equal to 1 if white	−0.5567**	−0.6132**	−0.6060***
	(−2.30)	(−2.50)	(−2.74)
Binary equal to 1 if married			−0.3137**
			(−2.01)
Past Activities			
Binary equal to 1 if first arrest a serious personal crime		1.3230***	0.9235***
		(3.92)	(2.99)
Binary equal to 1 if first arrest a less serious personal crime		0.0861	0.2220
		(0.29)	(0.85)

Table 8.1 Continued

Binary equal to 1 if first arrest a property offence		0.1824 (0.50)	0.1317 (0.49)
Number of times in police custody as a juvenile		0.1290*** (3.09)	0.1171*** (3.04)
Percentage of juvenile police contacts resulting in formal criminal justice processing		−0.0028 (−0.71)	−0.0024 (−0.70)
Binary equal to 1 if gang member		0.6968*** (3.18)	0.7035*** (3.97)
Constant	1.2440 (0.62)	0.5466 (0.26)	0.2476 (0.12)
Variance of estimated individual effects	0.9534*** (11.06)	0.8683*** (6.79)	0.6668*** (5.86)
Log likelihood	−445.52	−420.45	−410.94
n	2856	2856	2856

* Significant at the 0.10 level, two-tailed test.
** Significant at the 0.05 level, two-tailed test.
*** Significant at the 0.01 level, two-tailed test.

(e.g. police resources, family background and neighbourhood characteristics). The second column contains results for a specification that also includes predetermined variables related to the juvenile criminal record. The last column is for a specification including variables related to activities that occurred in the current year (e.g. fraction of the year employed) or previous years, possibly during the sample period (e.g. high-school graduation). We estimate three specifications as a partial check of the robustness of results. The implications of the Tobit models are similar to those of the probit model, the estimated coefficients are not reported. The results are available in Tauchen, Witte and Griesinger (1988).

The probit and Tobit models are significant in all specifications and the estimated coefficients, when significant, are of the same sign in all models. The estimated coefficients on the variables of primary interest are stable in sign and magnitude across specifications for a given estimation technique. We find greater time working, greater time in school and higher IQ to be significantly related to lower probabilities of criminal activity. The receipt of a high-school diploma, however, has no significant effect on offending for any specification or panel. Upon considering only the results for the time allocated to work and IQ, we could interpret our findings as indicating that a lower level of criminal activity is associated with greater income.[10] In light of the results for the high-school degree binary and the time allocated to school, the interpretation is less clear-cut.

The coefficients on the proportion of time working and the proportion of time in school are not significantly different from one another. Other studies (e.g. Farrington *et al.*, 1986; Gottfredson, 1985; Viscusi, 1986a) obtain the same results.[11] The standard economic explanation for this finding is based on a dynamic, human capital model of criminal behaviour (e.g. Flinn, 1986). Current employment and schooling could have similar effects since both affect permanent income. Employment has an obvious, direct bearing on current and permanent income. Schooling influences permanent income through its human capital effects on future wages and employment prospects.

Our findings for high-school graduation, however, do not indicate significant human capital effects on crime. Nor can the insignificant coefficients on the high-school graduation variable be explained by collinearity. Also, other researchers (e.g. Schmidt and Witte, 1984) report that wage rates are not significantly related to criminal activity.

Attending a parochial high school is significantly related to lower levels of criminal activity in the young adult years (18–25 years of age). Parochial schools have been found, in general, to do a better job of improving educational achievement, maintaining good attendance and reducing misbehaviour[12] than do public schools in cities such as Philadelphia. The human capital effects of parochial school attendance do not, however, lead to higher incomes. Kessler (1990) reports that graduates of parochial schools are more likely to have white-collar jobs than are public school graduates, but that there is no significant difference in their wages, *ceteris paribus*.

The results of other researchers and this study might be consistent with a model of criminal activity that conceives of legitimate time uses and social associations (e.g. participation in church activities, white-collar employment) as shaping or revealing preferences concerning illegal activities. Such a model might be developed by allowing the parameters of the utility function to depend on how an individual uses time or on an individual's associates (Theil, 1980; Phlips, 1983; Becker, 1992).[13] Note that this approach would not be inconsistent with standard economic models. Further, such an approach might address the criticism that economic models of crime are not applicable for teenage offenders, particularly young teenagers, few of whom are in the labour market (Felson, 1993).

The negative coefficient on the binary for whether the individual was white[14] is consistent with the common finding that non-whites have far higher crime participation rates than whites (e.g. Blumstein *et al.*, 1986). In our sample, though, this finding could be partly attributable to characteristics of the criminal justice system. A study that uses the same data set (Collins, 1985) reports that blacks were more likely to be arrested given the crime committed than were whites. Since we use arrest to measure criminal activity, the estimated coefficient on the binary for whether the individual was white reflects differences in police arrest practices as well as differences in criminal behaviours across racial groups. As in other studies that use official crime data, we find that the probability of arrest is higher for young men whose household head during high school had a relatively low-status occupation. In light of this result it may seem surprising that young men who grew up in higher-income neighbourhoods were more likely to be arrested in their early adult years. This is a common result, however, in studies that use neighbourhood or other aggregate measures of income. In previous studies, researchers interpret the average community income as measuring the opportunities for crime, and often find this variable to be positively related to the crime rate.[15] The findings related to the young men's juvenile criminal record and to the other family background variables are also consistent with the previous literature.

We find robust and significant general deterrent effects from greater criminal justice system resources. For the results reported in Table 8.1 the criminal justice resource variable is measured as real police resources per offence for the city of Philadelphia, and the estimated coefficient is negative and significant. We also find significant deterrent effects for the following measures of criminal justice resources: police officers per offence; total criminal justice employees (police, courts, and local corrections) per offence; real police budget per young male; and real police budget per capita. In considering these general deterrence results keep in mind that our measure of crime is arrests. The net impact of increased criminal justice resources on arrests is the sum of two opposing effects. Increasing criminal justice resources may have a deterrent effect on the level of criminal activity but also leads to a higher probability of arrest for any given crime. The estimated coefficients on the criminal justice resource variables therefore understate the pure general deterrent effect.

VI Conclusions

As in previous studies,[16] our findings provide little evidence that wages or incomes have consistently significant effects on crime. Our work does add to the growing literature that finds both employment and school attendance to be significantly related, in virtually identical ways, to lower levels of criminal activity. Similar effects for employment and schooling are difficult to explain using either a static crime as work model or the static demand-type model developed in Section III. Perhaps different types of criminal models will offer greater insights regarding the effect of legitimate time uses on criminality. Researchers have begun to develop dynamic models of criminal activity (e.g. Flinn, 1986) and models of crime based on psychological and sociological processes (e.g. Akerlof and Dickens, 1982; Dickens, 1986; Lattimore and Witte, 1986; Lattimore, Witte and Baker, 1992). Gary Becker is currently developing models for the 'rational formation' of preferences based on habitual behaviour and peer group influences (Becker, 1992). This work, like Becker's 1968 article, may change how researchers model criminal activity.

Notes

1   For a survey see Heineke (1978) or Schmidt and Witte (1984).
2   Some economists (e.g. Cook and Zarkin, 1985) have estimated statistical time-series models.
3   See Blumstein *et al.* (1978), Brier and Fienberg (1980), Cook (1980), Long and Witte (1981), or Freeman (1986) for surveys.
4   Good, Pirog-Good and Sickles (1986) use a police policy variable that might under our model be interpreted as a general deterrence variable. However, neither their model nor empirical results are consistent with such an interpretation.
5   Erez (1987: 132) provides a good survey of the research on how offenders approach crime. She concludes that 'criminal violations are mostly situational and that impulsive crime is more common than planned crime'.
6   Selling drugs may be an exception. See Freeman (1991) for a recent survey of the income of drug dealers.
7   See Tauchen, Witte and Long (1991) or Blumstein *et al.* (1983) for discussions of the general determinants of criminal justice system actions.
8   The notation with multiple arrests is messy and complicated. The implications of the model for structuring the empirical work are the same as for the above model.
9   Members of the sample had 147 arrests during the sample period. Eight per cent of the arrests were for crimes with potential or actual violence (homicide, rape, assault, and robbery); 25 per cent involved theft of property (burglary, larceny and motor vehicle theft) and the remainder involved offences such as drug sales or possession and buying or receiving stolen property.
10 The negative coefficient on the IQ variable might also arise because more intelligent individuals are better able to elude arrest for their criminal acts than are others.
11 As Viscusi (1986a) and others have pointed out, the coefficients on variables such as employment and schooling must be interpreted with care. In a standard human capital model these coefficients might be regarded as partial correlations with the crime variable.

12   These insights are drawn from Wilson and Herrnstein (1985) who provide a good survey of the effect of schools on criminality.
13   In dynamic models there are other possibilities. For example, one might incorporate state dependence (for a discussion of a possible approach see Heckman, 1981) or the type of 'taste shifter' vector discussed by MaCurdy (1985). In an interesting discussion of dynamic demand systems, Phlips (1983: 178) suggests that taste changes may 'result from better outside information due to external influences on a consumer, or they are of the "built-in" type, being related to past decisions'.
14   The three Hispanics in our sample were classified as non-white.
15   See Long and Witte (1981) for a review of this literature.
16   See Long and Witte (1981) and Freeman (1986) for reviews.

References

Akerlof, G. A. and Dickens, W. T. (1982) 'The Economic Consequences of Cognitive Dissonance', *American Economic Review*, 72, pp. 307–19.
Becker, G. (1968) 'Crime and Punishment: An Economic Approach', *Journal of Political Economy*, 76, pp. 169–217.
— (1992) 'Habits, Addictions and Traditions', *Kyklos*, 45, pp. 327–42.
Blumstein, A., Cohen, J., Martin, S. E. and Tonry, M. H. (eds) (1983) *Research on Sentencing*, 2 vols (Washington, DC: National Academy Press).
— Cohen, J. and Nagin, D. (eds) (1978) *Deterrence and Incapacitation* (Washington, DC: National Academy of Sciences).
— Cohen, J., Roth, J. A. and Visher, C. A. (eds) (1986) *Criminal Careers and 'Career Criminals'*, 2 vols (Washington, DC: National Academy Press).
Brier, S. S. and Fienberg, S. E. (1980) 'Recent Econometric Modeling of Crime and Punishment', *Evaluation Review*, 4, pp. 147–91.
Chamberlain, G. (1984) 'Panel Data', in Z. Griliches and M. Intriligator (eds), *Handbook of Econometrics*, vol. 2 (Amsterdam: North-Holland) pp. 1247–317.
Collins, J. J. (1985) 'The Disposition of Adult Arrests', unpublished manuscript, Wharton School, University of Pennsylvania.
Cook, P. J. (1979) 'The Clearance Rate as a Measure of Criminal Justice System Effectiveness', *Journal of Public Economics*, 11, pp. 135–42.
— (1980) 'Research in Criminal Deterrence : Laying the Groundwork for the Second Decade', in N. Morris and M. Tonry (eds), *Crime and Justice: An Annual Review of Research*, vol. 2 (Chicago: University of Chicago Press), pp. 211–68.
— and Zarkin, G. A. (1985) 'Crime and the Business Cycle', *Journal of Legal Studies*, 14, pp. 115–28.
Crowley, J. E. (1981) 'Delinquency and Employment', Working Paper, Center for Human Resource Research, Ohio State University.
Dickens, W. T. (1986) 'Crime and Punishment Again', *Journal of Public Economics*, 30, pp. 97–107.
Ehrlich, I. (1973) 'Participation in Illegitimate Activities: a Theoretical and Empirical Investigation', *Journal of Political Economy*, 81, pp. 521–65.
Erez, E. (1987) 'Situational or Planned Crime and the Criminal Career', in M. E. Wolfgang *et al.* (eds), *From Boy to Man: From Delinquency to Crime* (Chicago, IL: University of Chicago Press), pp. 122–33.

Farrington, D. P., Gallagher, B., Morley, L., St Ledger, R. J. and West, D. J. (1986) 'Unemployment, School Leaving and Crime', *British Journal of Criminology*, 26, pp. 335–56.

Felson, M. (1993) 'Social Indicators of Criminology', *Journal of Research in Crime and Delinquency*, 30, pp. 400–11.

Flinn, C. (1986) 'Dynamic Models of Criminal Careers', in A. Blumstein *et al.* (ed.), *Criminal Careers and Career Criminals*, vol. 2 (Washington, DC: National Academy Press), pp. 356–79.

Freeman, R. B. (1986) 'Who Escapes? The Relation of Churchgoing and Other Background Factors to the Socioeconomic Performance of Black Male Youths from Inner-City Tracts', in R. B. Freeman and H. J. Holzer (eds), *The Black Youth Employment Crisis* (Chicago, IL: University of Chicago Press), pp. 353–79.

— (1991) 'Crime and the Employment of Disadvantaged Youths'. Working Paper No. 3875, National Bureau of Economic Research, Cambridge, MA.

Good, D. H., Pirog-Good, M. and Sickles, R. C. (1986) 'An Analysis of Youth Crime and Employment Patterns', *Journal of Quantitative Criminology*, 2, pp. 219–36.

Gottfredson, D. C. (1985) 'Youth Employment, Crime and Schooling', *Development Psychology*, 21, pp. 419–32.

Heckman, J. J. (1981) 'Statistical Models for Discrete Panel Data', in C. F. Manski and D. F. McFadden (eds), *Structural Analysis of Discrete Data with Econometric Applications* (Cambridge, MA: MIT Press), pp. 114–78.

Heineke, J. M. (1978) 'Economic Models of Criminal Behavior: An Overview', in J. M. Heineke (ed.), *Economic Models of Criminal Behavior* (Amsterdam: North-Holland), pp. 1–33.

Hirschi, T. (1969) *Causes of Delinquency* (Berkeley, CA: University of California Press).

Holzman, H. R. (1982) 'The Serious Habitual Property Offender as "Moonlighter" ', *Journal of Criminal Law and Criminology*, 73, pp. 1774–93.

Karni, E. and Schneidler, D. (1990) 'Fixed Preferences and Changing Tastes', *American Economic Review*, 80, pp. 262–67.

Kessler, D. (1990) 'Are Private Schools Better at Training Students or Are They Merely Training Better Students? The Public–Private Achievement Differential'. Unpublished manuscript (MIT).

Lattimore, P. K. and Witte, A. D. (1986) 'Models of Decisionmaking under Uncertainty', in D. B. Cornish and R. V. Clarke (eds), *The Reasoning Criminal: Rational Choice Perspectives on Offending* (New York: Springer-Verlag), pp. 129–55.

— Witte, A. D. and Baker, J. (1992) 'The Influence of Probability on Risky Choice: a Parameteric Examination', *Journal of Economic Behavior and Organisation*, 17, pp. 377–400.

Long, S. K. and Witte, A. D. (1981) 'Current Economic Trends', in K. N. Wright (ed.), *Crime and Criminal Justice in a Declining Economy* (Cambridge, MA: Oelgeschlager, Gunn & Hain), pp. 69–143.

MaCurdy, T. E. (1985) 'Interpreting Empirical Models of Labor Supply in an Intertemporal Framework with Uncertainty', in J. Heckman and B. Singer (eds), *Longitudinal Analysis of Labor Market Data* (Cambridge: Cambridge University Press), pp. 111–55.

Montmarquette, C. and Nerlove, M. (1985) 'Deterrence and Delinquency', *Journal of Quantitative Criminology*, 1, pp. 37–58.

Myers, S. (1983) 'Estimating the Economic Model of Crime', *Quarterly Journal of Economics*, 98, pp. 157–66.

Petersilia, J., Greenwood, P. and Lavin, M. (1977) *Criminal Careers of Habitual Felons* (Santa Monica, CA: Rand Corporation).

Phillips, L. and Votey, H. L. (1984) 'Black Women, Economic Disadvantage and Incentives to Crime', *American Economic Review*, 74, pp. 293–7.

Phlips, L. (1983) *Applied Consumption Analysis* (Amsterdam: North-Holland).

Poterba, J. M. (1987) 'Tax Evasion and Capital Gains Taxation', *American Economic Review*, 77, pp. 234–39.

Rossi, P. H., Berk, R. A. and Lenihan, K. J. (1980) *Money, Work and Crime: Experimental Evidence* (New York: Academic Press).

Schmidt, P. and Witte, A. D. (1984) *An Economic Analysis of Crime and Justice* (New York: Academic Press).

Sellin, T. and Wolfgang, M. (1964) *The Measurement of Delinquency* (New York: John Wiley).

Swanson, G. and Tabbush, V. (1988) 'The Economics of Crime and Punishment', *Contemporary Economic Issues*, pp. 1–6.

Tauchen, H., Witte, A. D. and Griesinger, H. (1988) 'Deterrence, Work and Crime'. Working Paper no. 2508, National Bureau of Economic Research, Cambridge, MA.

—, Witte, A. D. and Griesinger, H. (1994) 'Criminal Deterrence: Revisiting the Issue with a Birth Cohort', *Review of Economics and Statistics*, 76, pp. 399–412.

—, Witte, A. D. and Long, S. K. (1991) 'Domestic Violence: A Nonrandom Affair', *International Economic Review*, 32, pp. 491–511.

Theil, H. (1980) *The System-Wide Approach to Microeconomics* (Chicago, IL: University of Chicago Press).

Thornberry, T. P. and Christenson, R. L. (1984) 'Unemployment and Criminal Involvement', *American Sociological Review*, 49, pp. 398–411.

Viscusi, W. K. (1986a) 'Market Incentives for Criminal Behavior', in R. B. Freeman and H. J. Holzer (eds), *The Black Youth Employment Crisis* (Chicago, IL: University of Chicago Press), pp. 301–46.

— (1986b) 'The Risks and Rewards of Criminal Activity: a Comprehensive Test of Criminal Deterrence', *Journal of Labor Economics*, 4, pp. 317–40.

Wilson, J. Q. and Herrnstein, R. J. (1985) *Crime and Human Nature* (New York: Simon & Schuster).

Witte, A. D. (1980) 'Estimating the Economic Model of Crime with Individual Data', *Quarterly Journal of Economics*, 94, pp. 57–87.

9

Editors' Introduction

Like Witte and Tauchen, Gerald Prein and Lydia Seus seek to understand the relationship between crime and economic factors by a panel study design, but using event history analysis to study the causal dynamics of the relationship. They place their work in the context of social exclusion, which can be construed as a denial of the 'opportunities' central to the economic calculus. Prein and Seus define a set of ordinal models for violent offences and property offences committed by young people and test them against data from a four-wave panel study in which several hundred male and female graduates of German vocational education programmes participated. The dynamic, longitudinal perspective leads Prein and Seus to focus on changes in the risk of committing delinquent offences at different stages of the life-course. They find no evidence for the hypothesis that failure to complete the vocational apprenticeship, or experience of unemployment, leads to delinquency. Rather, they find evidence that a cumulation of experiences of social exclusion can occur so that labour market participation and contacts with criminal justice agencies reinforce each other. For males, positive employment status served as a substantial mitigator when called to account by criminal justice agencies, a significant point in itself but especially so in that criminal offending by those in steady, vocationally qualified employment exceeded that by the unemployed. For females, unemployment was discounted as a criminogenic factor because they had recourse to traditional home-making and reproductive roles, an alternative to vocationally qualified work which was often strongly encouraged by partner and family.

'The Devil Finds Work for Idle Hands to Do': the Relationship between Unemployment and Delinquency*

Gerald Prein and Lydia Seus

Introduction

The issue of juvenile delinquency and, in particular, the presumed rise of violent crimes committed by young offenders, has resurfaced in the public debate in Germany. Phenomena of social insecurity are generally regarded as consequences of social crises and the decline of the integrative force of central institutions. Based on these assumptions, different policies are suggested. On the one hand, *zero tolerance*, i.e. the return to the imposition of rigid sanctions by the judicial system. On the other hand, calls for a better integration of juveniles into society by providing work and training opportunities in order to prevent delinquency. Successful vocational training and employment stability, which are still regarded as the 'normal situation' of occupational careers (Osterland, 1990), are considered one of the most important mechanisms to prevent social exclusion. To achieve this at a time when the concept of 'employment for all' is in crisis is proving more and more difficult.

Phenomena linked to processes of social exclusion are used in various theories explaining deviance and delinquency. Major themes in the debate on juvenile delinquency are 'poverty and delinquency', 'socialisation deficits and deviance', and 'unemployment and delinquency'.

The economic theory of crime (Becker, 1968; Ehrlich, 1973) is currently favoured by politicians over 'therapeutic approaches' (Lindenberg, 1998: 60). It is also one of the various theories that try to provide scientific support for the thesis that there is a causal link between unemployment and delinquency. In economic choice models, deterrence and employment are both regarded as inhibiting factors for delinquency: whereas deterrence

*Translated in discussion with the authors by Wolfgang Hübner and Katherine Bird.

increases the opportunity costs for delinquency, employment increases the possibility of an alternative legal income. On the basis of this type of model Witte and Tauchen (1994) concluded 'that working and going to school significantly decrease the probability of committing criminal acts' (Witte and Tauchen, 1994, abstract and Chapter 8 in this volume).

The results of empirical research are far from being coherent and unanimous in regard to these hypotheses. Fagan and Freeman state, referring to economic explanations: 'If we had well-substantiated estimates of the responsiveness of criminal behaviour to economic incentives these explanations could be definitely assessed. But the existing literature offers little clarity on this question' (Fagan and Freeman, 1997: 4). Nevertheless, theories defining 'crime' mainly as a problem of the 'underclass' present an astonishing stability. It is therefore important to address this problem again.

These considerations form the background for the design of the longitudinal studies conducted by Project A3, 'Employment History and Delinquency', of the Special Research Programme 186,[1] which investigates the life-courses of school-leavers from *Haupt-* and *Sonderschule*,[2] in Bremen. Between 1989 and 1997 a panel study using standardised instruments (four waves, $n = 426/370$) and in-depth interviews (five waves, $n = 60$) was conducted. The standardised data collected on the whole cohort covered, on the one hand, self-reported delinquency and, on the other hand, occupational qualifications, employment histories, family and partnerships, involvement in peer groups, and leisure activities. As a rule, information on delinquent behaviour was gathered on a yearly basis. In addition, an enquiry to the *Bundeszentralregister* (Central Federal Register of Criminal Judgements) provided information on the degree of officially recorded crime among the young people.

The investigated group is particularly vulnerable to failure during their occupational training and employment trajectories. A cross-sectional analysis of the fourth wave of standardised data shows that, seven years after leaving school, only one-third of the respondents had managed to find a qualified job. Around one-third of the young women and around 14 per cent of the young men were unemployed. If there is a relationship between unemployment and delinquency, then it should be possible to find it in this group.

Quantitative Analyses

Method, Operationalisation and Variables

Previous research into the relationship between unemployment and delinquency tends to fall into three categories: *studies at the aggregate level.*

Even if we can trust the basic data provided by the employment offices and the police, with this type of research design there is still a high risk of drawing false ecological conclusions. *Cross-sectional analyses* of micro-data may avoid the problem outlined above. However, since the temporal dimension is ignored it is not possible to determine the direction of causality. Although *panel data* (e.g. Witte and Tauchen, 1994) offer the means to control the temporal order, such analyses miss the mark when causal statements are tested since these are counterfactual statements (Holland, 1986). If the sentence *'A became delinquent because he was unemployed'* is viewed as a causal statement, then it also means that if *A had not lost his job* (ceteris paribus) *he would not have 'gone off the tracks'*. In this model a *change* in employment status does not explain a *change* in delinquent behaviour.

For this reason we chose an event history analysis approach (Blossfeld and Rohwer, 1995) in which the focus is on change. To operationalise the *dependent* process, 'changes in the risk of delinquency', we employed an ordinal categorisation and distinguished between four categories: no delinquency, one to four offences, five to ten offences and more than ten offences within the relevant time-frame.

In the formulation of the models we assumed that the factors that prevent young people (to an increasing extent) from tending to commit delinquent actions are not necessarily the same factors that encourage them to desist from committing such actions. Therefore, we estimated separate models for an increase or a decrease in delinquency. Furthermore, since it is known that different conditions are used to 'explain' different types of offences we distinguish (roughly) between *property offences; violent offences* (assault, grievous bodily harm, robbery, aggravated robbery, fights), drug offences, and a very heterogeneous remainder group of 'other offences' such as vandalism, payment avoidance and traffic offences. In this chapter the analysis of the *violent* offences will be presented. On the basis of these considerations we estimated a transition rate model for discrete time points with an ordinal 'destination state' that represents the conditional probability of a change in level of delinquency (increase or decrease given a controlled starting level).

In order to test whether, and if so, to what extent in situations of unemployment, there is a *(ceteris paribus)* raised increase and a reduced decrease in delinquency, *employment status* is a central covariate in the model. As with the vast majority of the other covariates that will not be discussed here in detail, such as gender, involvement in peer groups, age, living arrangement, prior delinquency and contact with the justice system, this was measured at the start of the time interval for which the relevant information on the frequency of delinquency was available. The reference category was 'in an apprenticeship leading to a recognised vocational qualification'. We then differentiated between 'not in gainful employment' (without responsibility for a household and children), 'responsible for a household and children', 'training schemes not leading to a vocational qualification', 'unqualified jobs', i.e. those for which no qualifications are required, its converse, 'qualified work', and 'other' (e.g. compulsory military or social service).

This is in accordance with the hypothesis that the current situation in life, and the options and restrictions that it provides, exert the strongest influence on delinquency. In order to test the extent of cumulative effects of unemployment, as postulated by Sampson and Laub (1997), an additional covariate, 'cumulative unemployment' (in years), was constructed. As with all the other time-dependent covariates this was calculated anew for each time-point at which employment status was calculated.

Results

When we consider the results of our preliminary analyses in the light of the methodological considerations above, it is necessary to differentiate between two questions:

1. Does unemployment lead to a higher likelihood of committing delinquent acts among young people? If this were so then the model parameters for cumulated unemployment and current unemployment must be clearly positive in the 'increase' model, and the parameters for apprenticeship and qualified work would have to be clearly negative.
2. Does unemployment lead to young people remaining delinquent? If this were so then in the model for a decrease in delinquency clearly negative parameters for cumulated unemployment and current unemployment would be expected, and clearly positive parameters for apprenticeship and qualified work.

If we examine the results in Table 9.1, there are a number of effects which bear comment. For women the increase is smaller and the decrease more likely than for men, even after controlling for the effects of occupation, partnership, involvement in a peer group, previous delinquency and contact with institutions of the criminal justice system. The analysis of the effect of current employment status, however, produced unstable estimates that only varied marginally from zero in a few cases.[3] For violent delinquency no stable effects are recognisable; it is not even possible to discern the direction of the parameter: although unemployed youths,

Table 9.1 Ordinal models for violent delinquency

Parameter		Increase		Decrease	
		Beta	p	Beta	p
Intercept	> 0 / < 1	−2.238	0.000	−0.887	0.048
	> 1 / < 2	−3.309	0.000	−0.239	0.601
	> 4 / < 5	−5.112	0.000	0.874	0.070
	> 10	−6.347	0.000	—	—
Gender	Women	−1.416	0.000	1.647	0.012
Time (age)		−0.335	0.000	0.426	0.011
Prior delinquency	Violence	0.061	0.001	−0.024	0.291
	Drugs	−0.005	0.669	0.035	0.013
	Property	−0.004	0.611	0.004	0.735
	Other	0.003	0.668	0.000	0.996
Contact with the	Dismissal by prosecutor	0.281	0.302	−0.166	0.754
justice system	Dismissal by judge	1.167	0.000	0.327	0.499
	Sentenced	1.101	0.011	−1.742	0.019
Peer groups		1.014	0.000	−0.837	0.030
Living arrangement	Not with partner or parents	0.206	0.465	−1.076	0.036
	With partner or spouse	−0.274	0.453	1.138	0.211
Employment status	Unemployed	−0.027	0.942	0.610	0.317
	Head of household	1.350	0.030	−1.202	0.315
	Training schemes	−0.075	0.864	0.674	0.580
	Unqualified jobs	0.048	0.880	0.335	0.544
	Qualified work	−0.511	0.202	−3.537	0.019
	Other	0.294	0.459	−0.056	0.932
Cumulative unemployment		0.200	0.397	−0.786	0.034

Source: Bremen Youth Panel on Employment and Delinquency, 1989–97.

ceteris paribus, have a higher rate of increase for violent offences in comparison with young people in training schemes, this increase is lower than that for young people in apprenticeships, in unqualified jobs and even housewives or househusbands. With this result it cannot be concluded that unemployment promotes violent delinquency. Turning to the process of decreasing delinquency, the unemployed fall together with the youths in training schemes as the group for which the risk of committing a violent offence falls most rapidly. However, the cumulation of experiences of exclusion from the labour market resulting from a continuing lack of employment seems to be a factor that delays or makes the exit from violent delinquency more difficult.[4]

From the analysis above it is clear that there are no provable effects that indicate the efficacy of current unemployment for directly promoting delinquency. On the contrary, the effects work strongly in the opposite direction. However, an effect of unemployment can be observed when long-lasting experiences of marginalisation from the labour market are involved.

Qualitative Analyses

Patterns of Social Interpretation of Unemployment and Delinquency

In this section, the results of the quantitative analysis will be extended and put into context by the findings from the qualitative part of the project, which was composed of biographical interviews and an analysis of records held by institutions of the criminal justice system. The reaction of young people to unemployment depends to a large extent on its perceived importance and relevance for their biographies. Certain coping strategies can, although labelled criminal by the criminal justice system, represent for the actors involved a conflict-resolution strategy, or a meaningful activity.

This understanding is expanded by the justification strategies employed by probation officers and judges who represent the institutional side, which is where the power to define deviance lies. Bringing all levels together promises a better understanding of the complex relationship between different exclusion mechanisms.

'I'm Just Like That, I Run Away': Coping Strategies of the Unemployed

If we wish to transcend the limited question of the direct relationship between unemployment or failure in the labour market and deviant behaviour/criminalisation, then it is important to take seriously the subjective interpretations of those affected.

The reactions of the unemployed young people or those who failed in the education and training system were multifarious. They ranged from resignation, adaptation and satisfaction to resistance and a complete dropping-out of the training and employment system. The fact that every type of process for regulating social action, be it the education and training system, the division of labour or criminal law, displays a gender-specific divergent structure, is also reflected in the strategies developed for coping with unemployment.

Although the so-called 'normal employment situation' (Osterland, 1990), usually interpreted as continuous, lifelong employment, no longer dominates social reality, as a norm it nevertheless forms the central focal point of the male biography – particularly among those who are excluded from it.

For men, unemployment was more strongly associated with existential fears. The danger of 'ending up on the street' made it harder for them to construct an 'accepted' or 'integrated' masculinity (Kersten, 1995). They became very active in their search for a job in order to avoid this fate. This was particularly so among the young men who had dropped out of vocational training and had therefore not yet secured an occupational identity via their apprenticeship.

The high degree of acceptance of unqualified work as well as phases of unemployment was particularly striking. In the first few years after leaving school the young men were, on the whole, in agreement with the social status that resulted from their occupational position. At the same time their belief in the fair workings of a competitive society was unbroken, as is shown impressively by data from the quantitative survey. In the first wave, 94 per cent of respondents agreed with the item 'If I try hard enough I can achieve a great deal'; in the second wave it was 96 per cent, in the third wave 89 per cent and in the fourth wave still 85 per cent (Dietz *et al.*, 1997). It seems highly unlikely that within such a group a potential for resistance could develop that, as in the French suburbs, turns into protest which can be constructed as criminal behaviour. The reproduction of the underclass has, so far, been a smooth process. This was partly because the young adults accepted individual responsibility for their occupational failure. They still believed that with enough effort they could have improved their social status.

This attitude changed as the young adults aged and with the lengthening of periods without work and regular income.

Many of those affected increasingly resigned themselves to this fate, and some even developed serious psychosomatic illnesses:

> I had real problems with it [unemployment]. . . . I had lots of mental problems cos I'm the sort of bloke who needs to keep busy. . . . And then later I was in therapy. . . . I couldn't trust anyone cos I had nothing to do but think, and then I couldn't sleep. An' so, yeah, well, that all got to be too much. An' then my last girlfriend, that all broke up . . .
> (FjV-8)[5]

The lack of structure for everyday life and the increasing pressure arising from the expectations of parents, partners and friends became serious problems. Regardless of their subjective coping strategy for unemployment, both men and women suffered from their objectively precarious financial situation and their feeling of social isolation.

Female unemployment was nearly always the result of dropping out of, or failure during, the training phase. Many young women experienced their unemployment primarily as a relief. In view of the limited opportunities for them to secure an interesting, well-paid occupation with good prospects, many of the former *Hauptschule* graduates dropped out of their apprenticeships. The lack of opportunities forced them to rethink their plans for the future and their construction of femininity. Many of them withdrew into the private sphere, often supported in this decision by their partners.

In the context of pregnancy or motherhood the young women often accepted unemployment, or the restrictions imposed by their apprenticeship or its premature end.

This traditional construction of femininity was accompanied by a drop in, or complete break with, delinquency. Partners took on the role of 'cooling-out agents' in that they did not support their girlfriends' apprenticeships and oriented them towards a traditional picture of women.

The dominant coping strategies adopted by the women were acceptance, disappointment or resignation. Their strategies for resolving conflict tended to be passive ('I'm just like that, I run away', Ba1-23). On the whole, when the strategy for coping with unemployment was acceptance or relief, no hint of delinquency was found.

The example of the young women shows that not every form of marginalisation or exclusion from the labour market also leads to social exclusion. In the case of the former *Hauptschule* students who withdrew into the private sphere, a socially tolerated form of exit from the training and employment system is involved that is valid even if the withdrawal is premature or involuntary. The married woman's role as housewife and mother may not be fully recognised by society, but it protects from exclusion.

These 'soft' exclusions (Kronauer, 1995: 206) represent forms of social inequality. They place women in the effective informal control system of the private sphere. The construction of an adjusted, traditional femininity leads, on the one hand, to a high degree of conformity, but on the other hand leads to a reproduction of discrimination in both gender relations and the employment hierarchy. In describing this situation an extended conception of work is necessary: the women are not without work, but they are not in gainful employment, which has negative consequences for them. The concentration on the partner and the relinquishment of personal social advancement form the basis for durable dependency.

'If You've Got a Job, They Don't Send You To Jail So Fast': work as a Meaningful Category for the Production of Conformity?

Our finding that employment or occupational status was not a causal condition for conformity or deviance, contradicts everyday theories in which disadvantaged, deprived youths resort to illegal sources of income out of economic necessity.

Even if the postulated relationship between unemployment and deviance is not reflected in the empirical results, it still exists in everyday, pragmatic theories of personality and deviance.

Young men who had appeared in court believed that the judges acted in accordance with certain interpretation patterns:

> they think we're lazy, that we, you know, hang out all day, do nothing but get into trouble, they think. That judge, the one that I always had, he was just like that. Get a job, what do you do all day? Hang around and get into trouble, stuff like that.
>
> (TiI V-41)

> Well, I reckon, if you've got a job, they don't send you to jail so fast as if you ain't. Cos if you ain't, then you hang out on the street all day and get into trouble. But if you've got a job, if you're at work all day, then you ain't got so much time to get into trouble.
>
> (ToI V-58)

The belief is that there is an automatic mechanism linking lack of temporal, and therefore implicitly social, integration and delinquency. Committing criminal offences then merely becomes a question of opportunity.

The quantitative analysis showed, however, that it was the youths in an apprenticeship leading to a recognised qualification who had high levels of delinquency, and not the unemployed youths. This surprising result can be better understood if we take a look at the interviews; there we found a type of adaptation that we called the 'double life'.

On the one hand there was a successful employment trajectory, and on the other a high level of delinquency and contact with institutions of the criminal justice system. The youths in question had achieved direct entry to the vocational training system in the occupation of their choice without encountering any problems. Furthermore, they were willing to conform to expectations regarding performance and attitudes towards work:

Whatever happens I want to finish my apprenticeship, learn hard, pay attention, I want to get a good grade at the end.

(FgI-3)

The motivation to complete the apprenticeship at any price stemmed from the satisfaction with the chosen path and their plans for the future. Corresponding to their positive employment trajectory, these apprentices could fall back on a supportive social network in their private lives. The role of the family was primarily a caretaking and not a controlling one.

During the week the apprentices were conformist and highly motivated, and during their free time and at the weekend they sought fun and action, which frequently came in the form of potentially delinquent behaviour. Fun, action, thrills and deviant behaviour, primarily with the peer group, were the mechanisms employed by the young men to accentuate their youth and to distance themselves from adulthood. Delinquency among the young people in our sample strongly emphasised the aspect of youth. This behaviour was typical for both the successful and the unsuccessful. Delinquent acts not only served to compensate for failures at work or the frustration of unemployment, but they also represented action, fun, thrills; acts with which the young men could partially reject growing up.

How did this group, composed nearly entirely of men, manage to live on two tracks without sanctions imposed under criminal law endangering their occupational trajectory? We believe we have found the answer at the institutional level.

The young men's normal biography is usually primarily defined by their qualifications for and activities in the labour market. The work ethic is a key concept that, in the question of disciplinary measures, becomes the guiding concept for interpretations.

The institutions of social control and those responsible for education and training seem to agree that delinquency among youth should not be met with sanctions that are detrimental to them, if the youths in question appear to be good apprentices. The work ethic is produced by various regulatory mechanisms. In this area criminal law does not have a productive function, but a representational one (Cremer-Schäfer and Steinert, 1986). This means the construction of 'normality, of that which the individuals confront as expectations and behavioural requirements, is reflected in the lenses of deviance. Cremer-Schäfer and Steinert defined the concept as follows:

I define 'work ethic' by drawing on Moore's 'implicit social contract', as the legitimations and institutional arrangements that, in a society, define and determine who works, and why, how and under which conditions work should be conducted.

(Cremer-Schäfer and Steinert, 1986: 81).

This means that youths who fulfil the ideological expectation that regular work is the precondition for regular income are not criminalised so quickly by judges, both by diversion from formal disposal and by the application of juvenile instead of adult law.

'Who works, how and under which conditions work should be conducted' are, in the German social system, determined by schools and the qualifications they offer.

The traditional path for a male graduate of a *Hauptschule* is to learn a trade and through this achieve the status of a skilled craftsman. If such a person managed to establish a 'male normal biography' (Heinz, 1996) then the status 'good apprentice' or 'good worker' dominates, and, in spite of a high degree of delinquency, he can manage to avoid being labelled a 'social deviant'.

This view is supported by statements both from the interviews and from entries in the criminal records.

Yeah, I think if you've got a job when you're up in court then you come out of it better. (ToV-41)

In 1996 Tom appeared in court for the third time, charged with assault. The prosecution advocated repealing his probationary sentence, received, among other things, for two armed robberies in 1993, and adding an additional six-month custodial sentence, a total of 2½ years in prison. Tom is convinced that his job saved him from prison.

in December '96, I think, the trial was, and I'd already worked for two years at XY. The judge said, he now has a son, he's had a steady job for two years, he's going to, well, she wasn't going to send me to jail was she? . . . When it came to the sentencing, that's what she said. Then she set a fine of . . . DM and, hmmm, yeah an' then in January or whenever that was, I got the sack from XY [laughs]. So, I reckon if the trial had been in February or whenever, then – I dunno [*laughs*]. (ToV-38)

In the judgements the social worker's reports and the decisions to try cases in the juvenile court,[6] the employment status of the accused played an important role. However, as the following case illustrates, the inclusion of the category 'work' is more differentiated than theoretical considerations suggest.

In Alex's case the judge established the start of a destabilisation that, among other reasons, was based on the fact that Alex had no income.

[Alex] receives no form of financial support and, as he himself admits, has to be satisfied with what friends and acquaintances can spare. In reality, it is probable that he earns a living by committing crimes.

The lack of an apprenticeship or a job – 'he is without work and social ties' – also serves to legitimate a custodial sentence or a remand in custody. By the date of sentencing Alex had found accommodation and promised to start seeking a new apprenticeship:

I want to work again as an apprentice bricklayer and I have already started making plans for finding a position.

One month before the trial Alex started an apprenticeship as a bricklayer. The sentence was one year detention in a youth detention centre, suspended for three years. One reason for the decision to hear the 19-year-old's case in the juvenile court was his employment situation at the time the offence was committed.

Alex went to *Sonderschule*, he had not completed an apprenticeship and at the time of the offence he was homeless and destitute. This did not make him independent, but led to a complete destabilisation of the accused, which justifies the decision for the juvenile court.

The sentence was oriented towards his current employment situation:

Alex has now found a good position and is continuing his apprenticeship as a bricklayer. . . . The court has set probation at three years to make clear to the accused that, whatever may come, he should complete his apprenticeship so that he is capable of earning his own living.

The probation officer wrote:

During his apprenticeship as a bricklayer he [Alex] was repeatedly ill for the school-based theoretical training, which his firm interpreted as a lack of interest and consequently fired him. He is now looking for a new position and may soon start his compulsory military service in the army.

Three months later:

The employment situation is still not clear. In spite of all his efforts, the young man has not been able to find a follow-on apprenticeship as a cook. Therefore, he has now started to look for a regular job.

Eight months later Alex had not found a follow-on apprenticeship, he was working as a temporary labourer and at that time was unemployed. After a further four months the probation officer applied for a reduction of the sentence, which was granted with the justification:

The probationer was found guilty of many theft offences. During his time on probation he has made great efforts to secure an apprenticeship.

It was not the success of the efforts that was decisive, but the probationer's perceived attitude, his manifest efforts, and his recognition of the work ethic.

Further illustration of this point is provided by another case. A young man with a 'patchwork' employment biography was sentenced to detention for 18 months, suspended for two years. The reason given in the judgement was that:

It is to the credit of the accused that he worked frequently, that he always sought new employment even if he left after only a short time. . . . May this time help him to mature and assist him to regulate his personal and occupational plans so that he does not relapse into his old habits and crimes; so that one day this punishment can be dropped.

From these and other examples of the interim analysis of criminal records it becomes clear that an apprenticeship or regular employment has a positive effect on the sentence or social worker's reports; that they offer protection – especially for young men. More relevance is attached to the presumed work ethic of the accused or the probationer than to his actual employment status.

In the sentencing for criminal offences committed by youths the premature ending of an apprenticeship, or unemployment, can serve as mitigating circumstances and at the very least they justify the decision for a trial in juvenile court.

Therefore, our results highlight the relevance of the category 'employment', but question its unambiguous unilateral orientation.

Conclusion

The results of the quantitative study did not empirically confirm the aetiological thesis that failure in an apprenticeship or unemployment among youths and young adults leads to delinquency. However, the analysis of the standardised data indicated that a cumulation of experiences of exclusion can occur in the occupational biography and record of contacts with criminal justice agencies.

The qualitative analyses made clear that it is important to focus on the subjective ways of coping with different situations: both employment and unemployment can themselves be interpreted in different ways. The simple equations, 'unemployment implies a subjective lack of prospects' and 'employment implies success', become problematic when seen against the background of the heterogeneity of interpretations that we found.

The importance of gender as a principle of social structuration also emerges in this analysis: men and women are socially positioned and thereby unequally placed as gender groups along this dividing line. This has consequences for their employment status, their prospects, but also the form of social control to which they are subjected. Furthermore, due to the differences in their social positioning, men and women have different 'resources' when it comes to subjectively coping with unemployment.

Even though we reject a linear causality between unemployment and delinquency that does not mean that we consider the discussion of a possible connection as irrelevant. That the connection is more likely to be found at the level of normative, action-inspiring assumptions was shown by the qualitative analyses.

Although it became clear that, due to developments in the labour market, a 'normal employment situation' can no longer be seen as a generally valid pattern for life planning and lifestyle, it still has a normative function. Agencies of selection, from educators to judges, are, at least as far as young men are concerned, still oriented in their processes of definition and sanctions towards the key category of 'work ethic'. Due to the high degree of self-disciplining on the part of the young adults, the criminal justice system plays only a minor part in maintaining the 'work ethic'.

We can therefore conclude that it is not primarily the unemployed who cause social problems in terms of criminal behaviour. This means that a whole branch of research, that tried for a long time to prove the connection between poverty and delinquency, could explore a new perspective such as the connection between delinquency and wealth, between criminal behaviour and power. We think this direction is promising.

Postulating a causal connection between poverty and delinquency in order to highlight the situation of disadvantaged youth and young adults might have been well-meaning and probably still is. However, proving that there is no connection leads back to the root of the problem: are we willing to see the economic situation and lifetime perspectives of young people as a scandal if we cannot assume that there is any danger for society? We think that we should.

Notes

1 Project A3 is funded by the German National Research Foundation and has been running since the start of the Special Research Programme 186 at the University of Bremen in 1988. In addition to the authors the project is composed of Adreas Böttger, Beate Ehret, Fred Othold and Karl F. Schumann (Project Leader).
2 Secondary education in Germany is organised in a three-tier system. The least academically able pupils are sent to a *Hauptschule*, and those with special educational needs are sent to a *Sonderschule*.
3 Since the group under study does not constitute a random sample we cannot apply inference statistics. However, we do 'misuse' the standard error to indicate the stability (or instability) of estimates.
4 Similar effects have been estimated for other types of offences.
5 All interviewees received pseudonyms. FjV-8 means that the quote is from Fjordi during the fifth wave and appears on page 8.
6 In Germany, defendants aged between 18 and 21 may be tried in either the juvenile or adult court, depending on their personal degree of maturity (determined by the court). In almost all cases the trial takes place in the juvenile court.

References

Becker, G. (1968) 'Crime and Punishment: an Economic Approach', *Journal of Political Economy*, 76, pp. 169–217.
Blossfeld, H.-P. and Rohwer, G. (1995) *Techniques of Event History Modelling: New Approaches to Causal Analysis* (Mahwah: Erlbaum).
Cremer-Schäfer, H. and Steinert, H. (1986) 'Sozialstruktur und Kontrollpolitik. Einiges von dem, was wir glauben, seit Rusche und Kirchheimer dazugelernt zu haben', *Kriminologisches Journal*, 1 (1986) pp. 77–118.

Dietz, G.-U., Matt, E., Schumann, K. F. and Seus, L. (1997) *Lehre tut viel. Berufsbildung, Lebensplanung und Delinquenz bei Arbeiterjugendlichen* (Münster: Votum).

Ehrlich, I. (1973) 'Participation in Illegitimate Activities: a Theoretical and Empirical Investigation', *Journal of Political Economy*, LXXXI, pp. 521–65.

Fagan, J. and Freeman, R. B. (1997) 'Crime, Work, and Unemployment', *Crime and Justice: A Review of Research*, 25.

Freeman, R. B. (1995) 'The Labor Market', in J. Q. Wilson and J. Petersilia (eds), *Crime and Public Policy*, 2nd edn (San Francisco, CA: Institute for Contemporary Studies), pp. 171–91.

Heinz, W. R. (1996) 'Sozial Benachteiligung Jugendlicher und die individuelle Zuschreibung von Mißerfolg beim Übergang in den Arbeitsmarkt', in T. von Trotha (ed.), *Politischer Wandel, Gesellschaft und Kriminalitätsdiskurse* (Baden-Baden: Nomos), pp. 355–67.

Holland, P. (1986) 'Statistics and Causal Inference', *Journal of the American Statistical Association*, 81, pp. 945–60.

Kersten, J. (1995) 'Junge Männer und Gewalt', *Neue Kriminalpolitik*, 8, pp. 22–7.

Kronauer, M. (1995) 'Massenarbeitslosigkeit in Westeuropa: die Entstehung einer neuen "Underclass" ', in SOFI (ed.), *Im Zeichen des Umbruchs. Beiträge zu einer anderen Standortdebatte* (Opladen: Leske & Budrich), pp. 197–214.

Lindenberg, S. (1968) 'The Influence of Simplification on Explananda: Phenomenon-Centred versus Choice-Centred Theories in the Social Sciences', in H.-P. Blossfeld and G. Prein (eds), *Rational Choice Theory and Large-Scale Data Analysis* (Boulder, CO: Westview Press), pp. 54–69.

Osterland, M. (1995) ' "Normalbiographie" und "Normalarbeitsverhältnis" ', in P. Berger and S. Hradil (eds), *Lebenlage, Lebensläufe, Lebensstile, Soziale Welt*, vol. 7 (Göttingen: Domhauf), pp. 351–62.

Sampson, R. J. and Laub, J. H. (1997) 'A Life-Course Theory of Cumulative Disadvantage and the Stability of Delinquency', in T. Thornberry (ed.), *Developmental Theories of Crime and Delinquency* (New Brunswick, NJ: Transaction), pp. 113–16.

Witte, A. D. and Tauchen, H. (1994), 'Work and Crime: An Exploration Using Panel Data', Working Paper No. 4794 (Cambridge, MA: National Bureau of Economic Research), and Chapter 8 in this volume.

10

Editors' Introduction

Prein and Seus's findings imply that employment status does not influence crime desistance in the way we might predict using our general knowledge of the way that crime participation relates to age. Such unexpected findings are the benefit of research which offers more detailed data about the criminally active population than can be offered by official statistics. Similar motivation informed our attempt in the following chapter to derive a finer-grained understanding of the relationship between crime and economic factors by using time-series data from police force records and social data sets matched to police force region. We addressed income inequality and socioeconomic deprivation as well as unemployment, but in contrast to other studies did so using regionally disaggregated data in an attempt to determine if any regional differences appeared in the impact of economic factors on crime patterns. Assuming that economic factors were most likely to manifest themselves in crimes relating to material gain, we examined five types of theft: theft from a vehicle, shoplifting, 'other theft' (an official residual category), handling stolen goods, and theft by employees. Growth in unemployment showed a positive relationship to the first four types of theft. In that we sought to combine the economic model with indicators featuring in sociological analysis, we draw on these findings to discuss the economic model of crime and some wider methodological concerns.

Crime, Unemployment and Deprivation

Alan Clarke, Nigel G. Fielding and Robert Witt

The relationship between economic circumstances and levels of criminal activity has been the subject of much research. This chapter reviews the findings of major aggregate empirical studies of the influences of poverty, income inequality and unemployment on recorded crime rates in the USA and the UK, and presents findings from our study of regional variation in criminal offending in England and Wales over the period 1979–93, bringing jointly to bear the perspectives of economics and sociology. A feature of our analysis is the disaggregation of theft into five different offences: burglary, theft from a vehicle, shoplifting, other theft and robbery. In discussing our findings from the perspective of economic rational choice theory, attention is drawn to conceptual issues and methodological problems in drawing inferences from aggregate data, giving rise to a critical examination of the unmodified rational choice perspective.

Empirical Studies of the Relationship between Economic Factors and Crime Rates

Many contemporary macro-level theories of criminal behaviour and empirical studies of crime rates address the relationship between economic factors and crime. Although quantitative studies have grown in number recently, an association between the level of crime and economic conditions has long been mooted (Ferri, 1881; Bonger, 1916; Thomas, 1925). Poverty, income inequality and unemployment are the indicators most frequently used when exploring the relationship between economic conditions and crime. However, research findings are far from consistent.

A body of research in the USA on the social ecology of crime has focused on poverty and social disadvantage, using standard metropolitan statistical areas, cities or urban neighbourhoods as units of analysis. This research is inconclusive, particularly regarding crimes of violence. While some studies find high homicide rates in poor, socially disadvantaged areas (Curtis, 1974; Loftin and Hill, 1974; Williams, 1984; Loftin and Parker; 1985), others show low homicide rates associated with poverty (Messner, 1983). In research on crime rates and aggregate economic conditions in over 50 residential areas, Patterson (1991) notes an association between percentage of low-income households in an area and level of violent crime. Krivo and Peterson (1996), in research based on Columbus, Ohio, record that the most disadvantaged areas generate particularly high levels of violence. However, Blau and Blau (1982), using data from the 125 largest metropolitan areas in North America, suggest it is not so much poverty that leads to an increase in violent crime, but socioeconomic inequalities.

Scrutiny of the literature reveals that when a single proxy variable is used to measure poverty its link to crime is invariably found to be weak and inconsistent, whereas when multiple indicators of poverty are used, a strong, unambiguous link is usually found (Loftin and Parker, 1985). Also, when poverty is included among a number of independent or control variables in structural economic models of criminal activity, its potential criminogenic properties are confirmed. In a study assessing empirically the influence of wage inequality on both violent crimes and property offences, using data from 28 major metropolitan areas from 1975 to 1990, Fowles and Merva (1996) report a strong, positive relationship between crime and levels of poverty, concluding that 'we can unequivocally say that the absolute level of poverty is an important structural covariate of all types of criminal activity' (1996: 179).

Research in the USA into the relationship between income inequality and crime rates also features inconsistent findings. For example, although some studies fail to find a significant positive relationship between income inequality and homicide rates (Messner and Tardiff 1986), others report a tendency for homicide rates to be higher in cities with higher levels of inequality (Land, McCall and Cohen, 1990). There are also mixed results regarding property-related crime. In a study of the 53 largest metropolitan areas, Danziger and Wheeler (1975) report a positive association between income inequality and crime rates for both robbery and burglary. Rosenfeld (1986) analysed data for 125 statistical metropolitan areas, but found no relationship between inequality and robbery. Patterson (1991) finds no real evidence that the degree of income inequality in an area relates to either household burglary or violent crime. Research using measures of wage inequality as opposed to income inequality found no link between inequality and property crimes, but strong evidence of a positive link between inequality and homicide (Fowles and Merva, 1996).

Much research on crime and social disadvantage has concentrated on estimating the effects of unemployment. Most studies have used aggregate-level data. American findings generally reveal levels of inconsistency similar to those found in exploring the impact of poverty or income inequality on crime rates. Fox (1978) finds no relationship between unemployment and crime over the period 1950–74. For the period 1940–73, Brenner (1978) reports a positive relationship between the unemployment and homicide rates. Land and Felson (1976) find no relationship between unemployment levels and incidence of violent crime but a weak, positive relationship with property crimes. Analysing data for the period 1947–77, Cohen, Felson and Land (1980) note that while unemployment rates have a weak negative effect for robbery, effects are strong and positive for burglary.

Reviews of research on the USA take varying lines. Whereas Fox (1978) questions the very existence of a relationship between unemployment and crime, Long and Witte (1981) report a positive but insignificant relationship, while Cantor and Land (1985) describe the relationship as weak and negative. Chiricos (1987) claims such findings have created a 'consensus of doubt' about the nature and existence of a causal link between unemployment and crime. Reviewing 63 American cross-sectional and time-series studies, Chiricos concludes that their evidence 'does favour the existence of a positive, frequently significant unemployment/crime relationship' (1987: 203).

Freeman (1994) considers research from the mid-1980s onwards, noting that although time-series studies show that the overall crime rate, and rates for specific offences such as burglary, are positively related to unemployment, results are modest and of limited value in explaining the upward trend in crime. On cross-sectional studies he concludes that their evidence 'continues to support a positive link between unemployment and crime and suggests that inequality may be an important contributor to crime' (1994: 15).

Empirical research into crime and unemployment in Britain, using aggregate level data, is not as extensive as that in the USA. British studies divide into two types: studies based on cross-sectional or panel data which model either total crime rates or broad crime categories, such as theft and burglary (Carr-Hill and Stern, 1979; Willis, 1983; Reilly and Witt, 1992, 1996) and time-series studies which model specific crime categories (Wolpin, 1978). As in the USA, these studies provide mixed and inconsistent findings. While Carr-Hill and Stern (1979) detected no independent effect of unemployment on the offence rate using census data from 1961 to 1966, Willis (1983), using crime data from police force areas, found positive and significant unemployment coefficients for both crimes of violence and theft. Examining data from 1984 to 1992, Orme (1994) finds insufficient evidence to suggest a consistent correlation between unemployment and recorded crime at police force level.

Cross-sectional data relating to the late 1970s reveal a rise in numbers of burglaries, thefts and robberies following an increase in the unemployment rate (Pyle, 1989). In contrast, Timbrell (1990) concludes that data for the early 1980s does not suggest unemployment was an independent factor in determining crime. While Scottish data suggest a significant relationship between crime and unemployment (Reilly and Witt, 1992), when Pyle and Deadman (1994) extended the data temporally, the relationship disappeared. Research based on an annual pooled time-series of 42 police force areas in England and Wales, covering the period 1980–91, found that growth in unemployment was positively correlated with burglary and theft (Reilly and Witt,1996).

Evidence of a positive relationship between unemployment and property crime features in numerous studies. Using data for England and Wales for the years 1949–84, Hale and Sabbagh (1991) find a link between unemployment and crimes of theft and burglary. Concentrating on domestic burglary from 1977 to 1990, Dickinson (1994) finds a clear association between unemployment and crime among young men. A strong and significant relationship between burglary rates and unemployment is also reported by Borooah and Collins (1995) following their analysis of annual data for the standard regions of Britain between 1983 and 1992.

Despite these recent positive findings, results of empirical research over the years suggest the relationship between unemployment and recorded crime is inconsistent. Tarling (1982), examining over 30 studies, found a small majority unable to detect a significant relationship between unemployment and crime. Reviewing British and North American research published between 1959 and 1985, Box reaches the tentative conclusion that 'existing time-series studies provide some support for the idea that unemployment and crime, particularly for younger males, are causally linked' (1987: 78) and 'on balance, the evidence from cross-sectional studies favours the hypothesis that unemployment and crime are related' (1987: 85).

The unemployment rate is not the only macro-level economic indicator used to explore the complex link between socioeconomic conditions and crime rates. Allan and Steffensmeier have criticised previous research for its 'inadequate conceptualization of labour market conditions' (1989: 109). Their study of youth and property crime in the USA considers not only the unemployment rate, but the quality of jobs available. Devine, Sheley and Dwayne-Smith (1988) maintain that inflation as well as unemployment should be taken into account when assessing 'economic distress'. Although acknowledging disagreement among economists on the distributional effects of inflation they argue that inflation, by creating distributional conflict, undermines confidence in existing institutional arrangements and structures.

Field (1990; summary in Chapter 6 in this volume) criticises many time-series studies for using the unemployment rate as sole economic indicator. Examining the relationship between economic activity and crime trends in England and Wales between 1950 and 1987, he asserts that economic factors, other than unemployment, have a significant role in explaining fluctuations in both property crime and crimes against the person. He concludes that 'once the effect of personal consumption is taken into account, no evidence emerged ... that unemployment adds anything extra to the explanation of any type of crime. The whole relation between the business cycle and crime appears to be encapsulated in that between personal consumption and crime' (1990: 7).

Debate surrounding the link between deprivation and crime follows a long-standing focus in sociological criminology on the conditions of ecological areas associated with crime. The unemployment rate has featured as the single most important causal variable in many empirical studies. Most make an underlying assumption that the causal relationship is one way: unemployment leads to crime. The fact that involvement in crime may precede unemployment is often ignored (see also Freeman, Chapter 6 in this volume). As Tarling (1982) observes, young people are likely to engage in crime long before they enter the labour market. Once 'embedded' in criminal environments their chances of securing legitimate adult employment are reduced (Hagan, 1993). Reciprocal causal influences may also be at work, with unemployment and crime mutually influencing one another throughout the individual's life course (Thornberry and Christenson, 1984).

Clearly there is scope to use other aggregate-level indicators in addition to the unemployment rate. Compounded social factors, economic circumstances and lifestyle make for a complex link between deprivation and crime. Our structural approach assumes that aggregate characteristics of geographical areas have independent effects on crime rates that are not attributable to the characteristics of individuals. We acknowledge a distinction between the linkage of economic circumstances and crime at macro-level, and the interaction between disadvantage and offending at individual level.

A Regional Analysis of Variation in Criminal Offending in England and Wales

Unlike many other empirical studies, our research into the impact of economic and socioeconomic variables on crime rates focused on *regionally disaggregated* data in order to detect any regional differences in the impact of economic factors. Our analysis is based on annual data from the ten standard regions of England and Wales: the North, North West, Yorkshire and Humberside, East Midlands, West Midlands, East Anglia, London, South East, South West and Wales. The study aimed to construct an annual time-series and regional cross-sectional data set of offences, unemployment, income variables, socioeconomic and sociodemographic data; assess the impact of economic disadvantage on rates of criminal offending over time; examine the character of regional variation in economic disadvantage and social malintegration, and to relate these differences to regional crime patterns.

Regional crime rates were constructed using data for the 42 police force areas in England and Wales. We concentrated on theft because of its frequency of occurrence and underlying economic motive; theft and handling stolen goods accounted for half of all crime recorded in England and Wales in 1993. Our analysis principally sought to estimate the effects of economic incentives to crime while controlling for changes in regional socioeconomic conditions. Regional unemployment rates and numbers of people in employment were obtained from Department of Education and Employment statistics. Income and wage data were derived from 'Economic Trends' and the New Earnings Survey. To obtain a measure of earnings inequality we compared the differential between regional manual gross weekly earnings at the 90th and 10th percentiles for male full-time employees. Measuring deprivation was more problematic, as there is no universally accepted definition. Following Townsend, Phillimore and Beattie (1988), three measures of relative deprivation by region were taken from the General Household Survey: percentage of people without access to a car, percentage living in rented accommodation and percentage in overcrowded accommodation. Table 10.1 provides summary statistics for unemployment rates and the categories of theft by region for selected years.

Table 10.1 first suggests that the pattern of regional unemployment rates in 1993 was similar to that in 1979. Second, regions with the highest unemployment rates record the highest crime rates. However, between 1979 and 1989 London experienced relatively low unemployment but remarkably high crime rates, particularly in theft from a vehicle and other

Table 10.1 Regional crime[1] and unemployment rates[2]

Region	Other theft			Theft from a vehicle			Shoplifting			Handling stolen goods			Unemployment		
	1979	1989	1993	1979	1989	1993	1979	1989	1993	1979	1989	1993	1979	1989	1993
North West	6.6	8.7	10.9	5.7	14.1	17.0	4.6	4.4	5.3	0.8	1.0	1.2	6.1	10.7	14.9
North	8.0	11.2	12.7	6.2	16.8	16.5	4.9	5.6	6.2	1.4	1.1	1.1	7.4	12.5	16.6
Yorkshire-Humberside	7.6	9.9	11.3	5.1	13.3	20.6	4.5	5.0	5.8	0.9	1.0	1.0	4.8	9.3	14.2
East Midlands	7.4	9.7	12.6	4.4	11.5	17.8	4.4	5.3	5.8	1.5	0.9	1.0	3.9	6.5	13.0
West Midlands	6.0	7.2	9.2	4.8	12.1	18.0	3.7	3.9	4.6	0.6	0.7	0.8	4.6	7.8	14.5
East Anglia	5.9	6.7	8.3	3.8	8.0	14.4	4.4	4.7	6.5	0.7	0.8	0.8	3.6	4.1	10.7
London	11.2	15.9	17.4	11.3	18.7	22.2	3.6	4.0	4.7	1.0	1.1	1.0	3.3	6.3	14.9
South East	5.4	6.8	9.1	4.1	8.4	16.0	4.1	4.1	5.6	0.8	0.7	0.9	3.0	3.5	12.5
South West	6.1	7.8	10.8	3.6	9.8	19.1	3.7	4.1	4.9	0.7	0.9	1.1	4.6	5.2	12.7
Wales	6.4	8.3	9.4	4.7	11.3	15.9	3.8	3.9	5.0	1.0	1.0	1.1	6.1	9.1	14.3

[1] Crime rates expressed per 1000 of population.
[2] Male unemployment rate.

theft. This suggests regional differences in the effect of unemployment on crime. Third, the regional pattern for 'other theft' in 1993 appears similar to that in 1979. Lower crime rates in this category are in the South East and East Anglia, with higher rates in London and the North. Fourth, relative regional crime rates for theft from a vehicle follow broadly similar patterns to those of 'other theft', ratings being lowest in East Anglia and the South East and highest in London and the North. Fifth, shoplifting slightly increased across all regions between 1979 and 1993. From 1979 to 1989 the highest rates occurred in the North and the East Midlands, the lowest in Wales, the West Midlands and London. The overall ranking shows little change during the early 1980s, although shoplifting in East Anglia rose to levels similar to those in the North. Finally, rates for handling stolen goods were relatively higher in the North and London with lower rates in the South East, West Midlands and East Anglia. In contrast to other crime categories, there has been little if any growth in this category between 1979 and 1993.

Detailed econometric analysis of the data (see Witt, Clarke and Fielding 1998a, 1999) shows that, first, high crime rates are associated with increases in male unemployment. Second, there is statistical evidence that changes in manual wages inequality are strongly positively correlated with changes in the crime rate, particularly for burglary, theft from a vehicle and robbery. Witt, Clarke and Fielding (1998b) give several reasons why crime rates might differ across regions. Using a common trend–common cycle approach to model the dynamics of English regional crime rates from 1975 to 1996, we provide evidence suggesting the existence of common trends and common cycles in regional crime rates.

Discussion

Although diverse findings from research into the relationship between aggregate economic conditions and crime rates have produced ambiguity at the theoretical level, most aetiological theories predict that when unemployment rises and/or income inequality increases, crime grows. Economics and sociology offer several theories supporting a link between unemployment and crime (Fagan, 1995). Most focus on the antecedents of crime in attempting to explain criminal motivation. Their primary objective is identification and explanation of economic conditions, structural factors and social forces likely to be criminogenic. As unemployment is essentially an economic variable we begin by viewing our findings from the perspective of economic rational choice theory.

Becker (1968 and Chapter 1 in this volume) presents a normative economic model of crime causation based on the concept of expected utility. This theoretical formulation assumes individuals are logical actors who make rational decisions to commit crime when expected benefits are seen to exceed perceived costs, making choices between legal activity (paid work) and illegal activity (crime) on their relative attractions. If paid employment becomes less rewarding, perhaps through increased earnings inequality, or crime appears to offer better rewards, then individuals are expected to invest less effort in legitimate activities and turn towards crime. For Becker, criminality is an occupational choice; individuals choose between mutually exclusive activities, namely, legal and illegal work. Individuals with limited labour market opportunities offering only poorly paid, insecure jobs, but with opportunity to obtain illicit rewards from crime, can be expected, under certain conditions, to choose the latter. It follows that the more disadvantaged the individual, the greater the potential gains from illegal relative to legal activity.

Utility maximization is also central to Ehrlich's (1973) version of economic choice theory, but decisions to engage in crime are not perceived as either/or choices. It is assumed that individuals can engage simultaneously in legitimate work and crime, or alternate between them as they experience intermittent unemployment. This allows for varying degrees of participation in illegitimate activities. Our data provide some support for an economic model of crime causation. Both growth in unemployment and increase in earnings inequality have a positive impact on property crimes. There is also evidence of a deterrent effect in the cases of theft from a vehicle and handling stolen goods.

While we would not make a full-blown critique of economic choice theory, we might highlight several concerns. There is a debate in criminology about the reliability and validity of official criminal statistics and the type of data these studies generally use. Unemployment and crime are 'contested concepts' that can be difficult to measure accurately. An observed rise in recorded crime may result from changes in police recording practices or victims' willingness to report crime, rather than 'real' growth in crime. British Crime Survey data suggest a decline in the proportion of reported offences that police actually record (Mayhew, Mirlees-Black and Maung, 1994). Reporting practices and procedures also vary between police forces. Empirical analyses based on official summary statistics are subject to problems of measurement error.

The aggregate nature of data sets is also important when interpreting empirical findings. It cannot be simply inferred from an observed ecological relationship between employment and crime that the unemployed are responsible for increases in crime rates. Ecological correlations between regional unemployment (or other deprivation indices) and regional crime rates say nothing about the relationship at individual level. Statistically significant correlations between recorded crime levels and unemployment figures provide no more than a basis on which to predict the type and levels of crime to be expected for given levels of unemployment (Jupp, 1989). To assume that relationships observed at aggregate level also apply at individual level falls foul of the 'ecological fallacy' (Robinson, 1950).

Herein lies a major problem with the economic choice approach: aggregated data are used to test microeconomic models predicated exclusively in terms of individual behaviour. Rational choice theory assumes individuals are rational decision-makers who engage in either legal or illegal activities depending on expected returns. Detection and punishment act as deterrents, being costs individuals take into account when making rational calculations and thus playing a role in determining criminal outcomes. These are reasonable theoretical assumptions in explaining how economic incentives feature in criminal aetiology but are best addressed by individual-level data. Only with information on the employment histories and offending behaviour of the same individuals can we determine the actual effect of unemployment on criminal motivation. Sequences and patterns of offending behaviour in relation to periods of unemployment reveal the more proximate causal relationships linking crime to unemployment. For example, a prospective individual-level panel study of 411 youths, combining longitudinal official crime and unemployment data with self-report data (Farrington *et al.*, 1986) found that more crime was committed by youths when unemployed than when in work.

Despite disagreement between theorists about how economic conditions influence individuals to engage in crime, it is accepted that 'at the aggregate level economic distress is ... a precipitator of negative social conditions that undermine legitimacy and order and weaken social bonds' (Devine *et al.*, 1988: 407). From the aggregate perspective the question is not who commits crime when unemployment rises, but what economic circumstances and social conditions are likely to have 'milieu' effects associated with high crime rates. While Cantor and Land (1985) maintain that increasing unemployment rates can raise a population's criminal motivational density, they assert that it does not follow that increased crime can be entirely accounted for by changes in motivational levels of the unemployed. Those in work are not immune to milieu effects.

To understand the causal mechanisms underlying the relationship between inequality and crime we must appreciate the wider sociocultural environment and social structural context in which criminality occurs. This can to an extent be provided by aggregate data. However, when moving beyond analysis of crime rates to address criminal propensity, macro-level data alone are insufficient; individual-level data are required. Aggregate studies may demonstrate marked inequalities in income and wealth distribution between socioeconomic groups. However, unless inequalities are perceived as unjust by those who experience them, and who consequently seek monetary rewards from crime, they cannot be assigned causal significance when seeking to establish a link between economic inequality and crime.

Note

The support of the Economic and Social Research Council is acknowledged. This chapter is based on a paper given at the British Criminology Conference (1997).

References

Allan, E. A. and Steffensmeier, D. J. (1989) 'Youth Underemployment and Property Crime: Differential Effects of Job Availability and Job Quality on Juvenile and Young Adult Arrest Rates', *American Sociological Review*, 54, pp. 107–23.

Becker, G. S. (1968) 'Crime and Punishment: an Economic Approach', *Journal of Political Economy*, 76, pp. 169–217.

Blau, J. R. and Blau, P. (1982) 'The Cost of Inequality: Metropolitan Structure and Violent Crime', *American Sociological Review*, 47, pp. 114–29.

Bonger, W. (1916) *Criminality and Economic Conditions* (Bloomington, IN: Indiana University Press).

Borooah, V. K. and Collins, O. (1995) 'Unemployment and Crime in the Regions of Britain: a Theoretical and Empirical Analysis', paper to the British Criminology Conference, Loughborough University.

Box, S. (1987) *Recession, Crime and Punishment* (London: Macmillan).

Brenner, H. M. (1978) 'Economic Crises and Crime', in L. Savitz and N. Johnston (eds), *Crimes in Society* (New York: Wiley), pp. 555–72.

Cantor, D. and Land, K. C. (1985) 'Unemployment and Crime Rates in the Post-World War II United States: a Theoretical and Empirical Analysis', *American Sociological Review*, 50, pp. 317–32.

Carr-Hill, R. A. and Stern, N. H. (1979) *Crime, the Police and Criminal Statistics: An Analysis of Official Statistics for England and Wales Using Econometric Methods* (London: Academic Press).

Chiricos, T. G. (1987) 'Rates of Crime and Unemployment: an Analysis of Aggregate Research Evidence', *Social Problems*, 34, pp. 187–212.

Cohen, L. E., Felson, M. and Land, K. C. (1980) 'Property Crime Rates in the United States: a Macrodynamic Analysis, 1947–1977', *American Journal of Sociology*, 86, pp. 90–118.

Curtis, L. A. (1974) *Criminal Violence* (Lexington, MA: D. C. Heath).

Danziger, S. and Wheeler, D. (1975) 'The Economics of Crime: Punishment or Income Redistribution', *Review of Social Economy*, 33, pp. 113–31.

Devine, J. A., Sheley, J. F. and Dwayne-Smith, M. (1988) 'Macroeconomic and Social Control Policy Influences on Crime Rate Changes, 1948–1985', *American Sociological Review*, 53, pp. 407–20.

Dickinson, D. (1994) 'Crime and Unemployment', Department of Applied Economics Working Paper, University of Cambridge.

Ehrlich, I. (1973) 'Participation in Illegitimate Activities: a Theoretical and Empirical Investigation', *Journal of Political Economy*, 81, pp. 521–65.

Fagan, J. (1995) 'Legal Work and Illegal Work: Crime, Work and Unemployment', in B. Weisbrod and J. Worthy (eds), *Dealing with Urban Crisis: Linking Research to Action* (Evanston, IL: Northwestern University Press), pp. 79–115.

Farrington, D., Gallagher, B., Morley, L., St. Ledger, R. and West, D. (1986) 'Unemployment, School Leaving and Crime', *British Journal of Crimonology*, 26(4), pp. 335–56.

Ferri, E. (1881) *Criminal Sociology* (Boston, MA: Little, Brown).

Field, S. (1990) *Trends in Crime and their Interpretation: A Study of Recorded Crime in Post-War England and Wales*, Home Office Research Study no. 119 (London: HMSO).

Fowles, R. and Merva, M. (1996) 'Wage Inequality and Criminal Activity: an Extreme Bounds Analysis for the United States, 1975–1990', *Criminology*, 34, pp. 163–82.

Fox, J. A. (1978) *Forecasting Crime Data* (Lexington, MA: Lexington Books).

Freeman, R. B. (1994) *Crime and the Job Market*, Working Paper No. 4910 (Cambridge, MA: National Bureau of Economic Research).

Hagan, J. (1993) 'The Social Embeddedness of Crime and Unemployment', *Criminology*, 31, pp. 465–91.

Hale, C. and Sabbagh, D. (1991) 'Testing the Relationship Between Unemployment and Crime: a Methodological Comment and Empirical Analysis Using Time-Series Data for England and Wales', *Journal of Research in Crime and Delinquency*, 28, pp. 400–17.

Jupp, V. (1989) *Methods of Criminological Research* (London: Unwin Hyman).

Krivo, L. J. and Peterson, R. D. (1996) 'Extremely Disadvantaged Neighborhoods and Urban Crime', *Social Forces*, 75, pp. 619–50.

Land, K. C. and Felson, M. (1976) 'A General Framework for Building Dynamic Macro-social Indicator Models Including an Analysis of Changes in Crime Rates and Police Expenditures', *American Journal of Sociology*, 82, pp. 565–604.

Land, K. C., McCall, P. L. and Cohen, L. E. (1990) 'Structural Covariates of Homicide Rates', *American Journal of Sociology*, 95, pp. 922–63.

Loftin, C. and Hill, R. H. (1974) 'Regional Subculture and Homicide', *American Sociological Review*, 39, pp. 714–24.

Loftin, C. and Parker, R. N. (1985) 'An Errors-in-variables Model of the Effect of Poverty on Urban Homicide Rates', *Criminology*, 23, pp. 269–85.

Long, S. K. and Witte, A. D. (1981) 'Current Economic Trends: Implications for Crime and Justice', in K. N. Wright (ed.), *Crime and Criminal Justice in a Declining Economy* (Cambridge, MA: Oelgeschlager, Gunn & Hain).

Mayhew, P., Mirlees-Black, C. and Maung, N. A. (1994) *Trends in Crime: Findings from the 1994 British Crime Survey*, Home Office Research and Statistics Department (London: Home Office).

Messner, S. F. (1983) 'Regional and Racial Effects on the Urban Homicide Rate: the Subculture of Violence Revisited', *American Journal of Sociology*, 88, pp. 997–1007.

Messner, S. F. and Tardiff, K. (1986) 'Economic Inequality and Levels of Homicide: an Analysis of Urban Neighborhoods', *Criminology*, 24, pp. 297–317.

Orme, J. (1994) 'A Study of the Relationship Between Unemployment and Recorded Crime', *Home Office Statistical Findings*, December, Issue 1/94 (London: Research and Statistics Department, Home Office).

Patterson, E. B. (1991) 'Poverty, Income Inequality and Community Crime Rates', *Criminology*, 29, pp. 755–76.

Pyle, D. J. (1989) 'The Economics of Crime in Britain', *Economic Affairs*, 9, pp. 6–9.

Pyle, D. J. and Deadman, D. (1994) 'Crime and Unemployment in Scotland', *Scottish Journal of Political Economy*, 41, pp. 325–35.

Reilly, B. and Witt, R. (1992) 'Regional Crime and Unemployment in Scotland: an Econometric Analysis', *Scottish Journal of Political Economy*, 39, pp. 213–28.

Reilly, B. and Witt, R. (1996) 'Crime, Deterrence and Unemployment in England and Wales: an Empirical Analysis', *Bulletin of Economic Research*, 48, pp. 29–51.

Robinson, W. S. (1950) 'Ecological Correlations and the Behaviour of Individuals', *American Sociological Review*, 15, pp. 351–7.

Rosenfeld, R. (1986) 'Urban Crime Rates: Effects of Inequality, Welfare, Dependency, Region and Race', in J. M. Byrne and R. J. Sampson (eds), *The Social Ecology of Crime* (New York: Springer), pp. 25–46.

Tarling, R. (1982) 'Crime and Unemployment', Home Office Research and Planning Unit Bulletin No. 12 (London: HMSO).

Thomas, D. S. (1925) *Social Aspects of the Business Cycle* (London: Routledge).

Thornberry, T. P. and Christenson, R. L. (1984) 'Unemployment and Criminal Involvement: an Investigation of Reciprocal Causal Structures', *American Sociological Review*, 49, pp. 398–411.

Timbrell, M. (1990) 'Does Unemployment Lead to Crime?', *Journal of Interdisciplinary Economics*, 3, pp. 223–42.

Townsend, P., Phillimore, P. and Beattie, A. (1988), *Health and Deprivation: Inequality and the North* (London: Croom Helm).

Williams, K. (1984) 'Economic Sources of Homicide: Re-estimating the Effects of Poverty and Inequality', *American Sociological Review*, 49, pp. 283–9.

Willis, K. G. (1983) 'Spatial Variations in Crime in England and Wales: Testing an Economic Model', *Regional Studies*, 17, pp. 261–72.

Witt, R., Clarke, A. and Fielding, N. (1998a) 'Crime, Earnings Inequality and Unemployment in England and Wales', *Applied Economics Letters*, 5, pp. 265–7.

—, —, — (1998b) 'Common Trends and Common Cycles in Regional Crime Rates', *Applied Economics*, 30, pp. 1407–12.

—, —, — (1999) 'Crime and Economic Activity: a Panel Data Approach', *British Journal of Criminology*, 39, pp. 391–400.

Wolpin, K. (1978) 'An Economic Model of Crime and Punishment in England and Wales, 1894–1967', *Journal of Political Eonomy*, 86, pp. 815–40.

Part III

Modelling the System-Wide Costs of Criminal Justice Policies and Programmes

11

Editors' Introduction

Economics has pursued an agenda in relation to crime which has been directed to understanding the part played by calculative action and allocative choice by offenders, assessing the effects of changes in the wider economic context against which they reach their decisions, and evaluating how criminal justice agencies should respond. However, criminal justice is a system, and economic analysis is helpful in monitoring the system-wide effects of resource allocation in any system. In this final section we offer examples of the contribution economic perspectives can make.

There has been much recent interest in assessing the true costs of crime. One motivation has been the attempt to inform policy decisions by a more discriminating assessment of the effects of different policies, which requires a close estimate of their costs. An example is the recent controversy over whether punishment 'works' and, specifically, the value of imprisonment relative to non-custodial measures. Another is the equally controversial matter of capital punishment where, again, cost–benefit analyses have been attempted. These are not the only matters where a perspective based on analysis of costs is worthwhile. It can be important, for example, in setting a sound basis for compensation claims by victims.

Lynch, Clear and Rasmussen review attempts to estimate the victim costs of crime, and offer refinements to existing techniques. Like the economic returns to crime, its cost to victims is not yet definitively established. Artful means are required to construct sufficiently precise estimates to take their place in formal models. Compared to surveys of victims or extrapolations from jury awards, Lynch, Clear and Rasmussen suggest that estimating the impact of crime on property values offers a useful alternative. The method is based on willingness to pay for reduced probability of victimisation, an approach with the precision of a cash nexus and based on the economist's orientation to marginal analysis, where costs of crime are modelled in terms of revealed preferences for improved safety (or risk reduction). Operationalisation of the model gives an interesting comparison between willingness to pay for lower risk of violent crime, indexed by house prices, and the actual cost of crime experienced by households. However, the price/risk trade-off behaves differently in high-crime areas, suggesting the operation of contagion and threshold effects in individual decisions that affect neighbourhood quality.

Modelling the Cost of Crime

Allen K. Lynch, Todd Clear and David W. Rasmussen

A growing literature attempts to estimate the victim costs of crime. Three approaches to modelling the cost of crime are examined in this chapter: surveys of victims, extrapolation from jury awards, and the impact of crime on property values. The first two methods attempt to measure the victim costs associated with specific crimes while the third is based on the willingness to pay for a reduced probability of victimisation. Section I summarises these three approaches to the cost of crime. Economics is oriented towards marginal analysis, which leads economists to model the cost of crime in terms of a revealed preference for small improvements in public safety. In Section II we present an empirical model of the willingness to pay for increased public safety which incorporates estimates of the victim costs of crime rather than a simple enumeration of the number of reported offences. Concluding comments follow in Section III.

I Three Approaches to Estimating the Cost of Crime

Asking crime victims about the costs they suffer is the most obvious way to obtain information about the costs of criminal activity. The US National Crime Victimization Survey (NCVS), for example, asks victims of crime about their economic losses, which include the value of stolen property, medical costs incurred, and forgone productivity as a result of the victimisation. This approach almost surely underestimates the victim costs of crime. For example, the NCVS asks if the respondent was victimised during the past six months, but some physical and psychological costs are likely to persist for a longer time period and, in some cases, do not become apparent as early as six months. Limiting the time frame has the advantage of eliciting a more accurate answer at the expense of ignoring longer-term costs. Victims may also underestimate total costs because they cannot accurately account for insured losses that do not involve out-of-pocket expenses. Of greater consequence, for violent crime at least, is that this type of survey does not adequately account for non-monetary psychological costs that are usually identified as 'pain and suffering'.

The second approach to estimating the cost of crime is to establish an empirical relationship between the direct costs of crime and the magnitude of jury awards in similar cases (Cohen, Miller and Weirsema, 1995). Juries can provide compensation to victims in the amount required, in the jury's judgement, to make the victim whole. Cohen, *et al.* (1995) estimate average pain and suffering for specific crime categories by investigating the relationship between pain and suffering costs awarded in jury trials and direct costs (medical costs plus lost wages). Their approach provides a more complete accounting of costs than the traditional survey methods because it includes short– and long-term productivity losses as well as pain and suffering.

Estimating pain and suffering in this way is controversial because it presumes that the relationship between direct costs and pain and suffering costs for all crimes can be estimated from what is in all likelihood a biased sample of cases that go to trial. The cases that are most likely to go to trial are those in which the plaintiffs and defendants disagree about the appropriate settlement. Since such disagreements are most likely when large monetary pain and suffering costs are claimed, the sample of cases going to trial might have greater than average pain and suffering costs relative to direct costs. Thus using jury awards to estimate average pain and suffering costs has been called 'voodoo economics' and 'bogus science' (Austin, 1996; Butterfield, 1996). The basic premise of this approach, however, is consistent with the contingent valuation method that is used in many areas that face the problem of measuring non-market benefits or costs. Widely used to assess environmental assets, the contingent valuation uses survey data to determine how much people are willing to pay to use or reserve the right to use an environmental asset such as a wilderness area. Jury awards can be considered a contingent valuation exercise using a small but very well-educated group of survey respondents to determine their willingness to accept compensation for all damages associated with the specific victimisation described in the legal proceedings.

Critics have also questioned the use of estimates derived from jury awards, in part because these pain and suffering costs do not reflect the expenditure of resources. In addition, there remains the question as to whether the resulting cost estimates are plausible. The enormously high pain and suffering estimates contribute the vast majority of the total cost of crime estimated by Cohen *et al.* (1995). This big gap between violent and other crimes also means that the vast proportion of total crime costs result from a small fraction of offences, making overall cost estimates volatile in relation to the rate of these crimes.

Despite these problems, attempts to incorporate pain and suffering in the crime–cost equation may be very useful. Lynch (1998) compares the cost estimates provided by Cohen *et al.* (1995) with those in the NCVS from the perspective of internal consistency, and finds the jury award method is preferred to the survey approach of NCVS because it provides more realistic estimates of the relative costs of violent crime. As a most compelling example, NCVS cost estimates show car theft to be more serious than rape, while Cohen *et al.* report rape to be 23 times more serious than car theft. Thus, while the specific cost estimates of the jury award method may be questioned, its principal advantage is that it provides more plausible and higher costs for violent crimes than for property crimes.

A third method of estimating the cost of crime is based on the proposition that there exists a relationship between crime levels and property values. The basic premise of this approach is that people are willing to pay a premium to live in a neighbourhood that reduces the risk of victimisation. Differences in the sales price of houses, when all other things are held constant, will capture the willingness to pay for a lower probability of victimisation. An implicit price of spatial variations in crime can therefore be estimated with implicit price (also called hedonic) models of housing markets.[1] Such models are intuitively appealing because they reflect the extent to which people 'vote with their feet' to obtain increased public safety.

The theoretical underpinnings of these models are well established. Goods with many attributes, such as housing or automobiles, can be decomposed into a bundle of underlying characteristics, each of which has an implicit price. The good's price reflects the amount of each characteristic times its implicit price. The selling price of a dwelling, for example, is determined by the characteristics of the dwelling (e.g. age and number of rooms), lot description (e.g., size and waterfront), neighborhood attributes (e.g. poverty rate) and the level and seriousness of crime in the area. Changes in any of these characteristics will affect a dwelling's selling price. Hedonic models are used to determine how changes in the underlying attributes – including factors such as crime, pollution and school quality – affect the market price of homes.

Thaler (1978) was the first to estimate the cost of crime with an implicit price model, using data from Rochester, NY, and found that the average property crime lowered house prices by approximately $1930 in 1995 prices. Studies by Hellman and Naroff (1979) and Rizzo (1979) used census tract data from Boston and Chicago respectively, and confirm that crime has a significant impact on house prices. Hellman and Naroff estimate that a 1 per cent change in the overall crime rate lowers house values by –0.63 per cent.

Cohen (1990) implies that these estimates are biased because they rely on reported crime rather than the cost of crime. He argues that using the number of crimes implicitly places the same 'value' on a larceny and a burglary, and treats an unarmed robbery as equivalent to a murder. Weighting crimes by their seriousness could result in a very different measure of public safety for jurisdictions with identical crime rates. Lynch and Rasmussen (1998) explore Cohen's conjecture by using Cohen *et al.'s* (1995) cost of crime estimates to weight the seriousness of individual crimes. The results support Cohen *et al.* The number of property crimes has a significant positive impact on house values in a hedonic model, while an alternative measure of the cost of property crime has the expected significant negative effect on house values. This result is expected if higher-income neighbourhoods are more likely than low-income areas to report numerous minor crimes.

II   Estimating the Cost of Crime with a Hedonic Model

The hedonic model is presented here as a preferred method of economists, because it provides estimates of the willingness to pay for marginal increments in public safety which here are measured by the victim cost of crimes. The data for this study come from the population of single-family dwellings sold in the Jacksonville (FL) housing market between 1 July, 1994 and 30 June, 1995. The selling price is regressed on house and lot characteristics, neighbourhood characteristics, and crimes occurring in the police beat in which the property is located:

$$P_i = f(S_i, N_i, C_k)$$

where
   P_i = the selling price of the home
   S_i = a vector of structural and lot characteristics
   N_i = a vector of neighborhood characteristics
   C_k = the estimated cost of reported crime in the relevant police beat.

Jacksonville (with a 1995 population of 712 000) offers several advantages for this study because the city and Duval county have a consolidated government. All housing units in the county therefore face one municipal government; one sheriff's department; the same local ordinances; and a common set of rules governing new housing developments, environmental regulations and building permit requirements. Since Florida has county-wide school districts, local variations due to different school board policies are also eliminated. Thus, the Jacksonville housing market can be examined in its entirety without as much concern for the myriad of policies that could affect housing prices in local jurisdictions within most US metropolitan areas.

Data on sale price and house and lot characteristics are from the Multiple Listing Service (MLS) and were provided by the Jacksonville Board of Realtors. The list of structural and lot characteristics included in the model comprises those customarily included in hedonic models and they are listed, together with descriptive statistics, in Table 11.1. The sample includes 2880 observations after excluding cases with missing information.

Moulton (1990) shows that the use of fixed neighbourhood definitions, such as census tracts, to gather neighborhood data will yield biased coefficients because many observations have common neighbourhood characteristics. Consequently we use a geographic information system to define a unique neighbourhood for each house in the sample. Using information from Equifax Data, a private vendor that provided estimates of economic and demographic characteristics for 1995 by block group, and employing MapBasic, a geographic information system programming language, data on the economic and demographic characteristics of the area immediately surrounding each observation were collected.

To gather data that are unique to each observation in the sample, the latitude and longitude of each house in the sample is found using a geographic coding program. A one-mile radial distance is swept around each observation to generate the neighbourhood characteristics. Because the information is available only at the block group level, these data are retrieved via 'proportional grabs'. Under this approach the neighbourhood includes all census block groups that are entirely within the circle as well as those that are partially included. In effect it is assumed that household characteristics are distributed evenly throughout the block groups, so characteristics of block groups that are partially in the circle are also included in the estimation of neighbourhood characteristics. These are the neighbourhood variables reported in Table 11.1.

The City of Jacksonville provided data on FBI index crimes (murder, manslaughter, rape, robbery, aggravated assault, burglary and larceny) for 87 police beats that cover the central city as well as suburban areas during 1994. Each house in our sample is placed into the appropriate police beat and then assigned a cost of crime measure which is created by weighting each crime committed by the victim cost estimates provided by Cohen *et al.* (1995).

Table 11.1 Definitions of variables used in housing market analysis

Variable definition	Mean	Standard deviation
Selling price of home	94 532.89	83 807.51
Log selling price of home	11.24412	0.630806
Structural variables		
Number of bedrooms	3.18826	0.657443
Number of bathrooms	1.975035	0.654404
Size of home (in square feet)	1738.56	662.7765
Age of the home	27.2282	19.77956
Log of age	2.9478	0.959611
Size of the lot (in acres)	0.320869	0.55176
Central heating unit dummy	0.849653	0.357474
Central air-conditioning dummy	0.830208	0.375515
Inground pool dummy	0.180271	0.38448
Waterfront property dummy	0.092361	0.289585
Assumable mortgage dummy	0.194444	0.395841
Gated community dummy	0.028472	0.166347
Number of covered parking places	1.430855	0.760911
Dummy for fenced property	0.744792	0.436054
Number of fireplaces	0.643403	0.556241
Dummy for vacant property	0.258681	0.437986
Neighbourhood variables		
Population	17543.09	13765.47
Proportion of population kids (17 and under)	0.227143	0.033428
Proportion of population white	0.799884	0.176985
Proportion of population own home	0.672649	0.142773
Proportion of population older (55 plus)	0.17308	0.063998
Proportion of population young adult (18–24)	0.132308	0.037568
Proportion of population Hispanic	0.038831	0.012178
Median household income ($)	304899.9	187338.7
Proportion of population white-collar employment	0.674559	0.110088
(Average–median income)^2 ($)	2.75E+08	4.72E+08
Population growth (1990–96)	0.067724	0.06074
Proportion of population bachelors degree	0.1621	0.077418
Crime-related variables		
All crimes in police beat	359.2417	273.0962
Property crimes in police beat	323.9104	252.7703
Violent crimes in police beat	35.33125	26.48515
Cost of all crimes in police beat ($)	2 175 976	2 511 364
Cost of property crimes in police beat ($)	367 271	305 153.5
Cost of violent crimes in police beat ($)	1 808 705	2 360 871

One might expect that most house sales occur in relatively safe police beats. Inspection of the data shows this to be true, but there is significant variation within the sample. Consider the distribution of observations by the number of murders. Almost 61 per cent of the sales in the sample were in police beats that had no murders reported. Fourteen per cent of the homes were exchanged in police beats with between one and 20 murders and about 25 per cent of the observations occur in beats with more than 20 homicides. Thus there is considerable variation in the amount of crime reported in proximity to the houses in our sample. To investigate the possibility that the crime coefficients do not adequately capture the impact of high crime on sales price, we rank-ordered the observations by the cost of crime and divided them into deciles. The regression reported in Table 11.2 includes a dummy variable for house sales in the ninth and tenth deciles and interaction variables that provide estimates of the extent to which house and neighbourhood characteristics in high-crime areas are valued differently than those in other areas.

Table 11.2 reports the results for regression models that show that the house characteristics and neighbourhood coefficients are generally in accord with *a-priori* expectations. Most property characteristics are significant at the 0.01 level. Older homes and those vacant at the time of sale are sold at a discount, as are those with more bedrooms. Although more bedrooms might be expected to have a positive coefficient, this result is expected because larger rooms are preferred to smaller ones for a house with a given number of square feet (Rasmussen and Zuehlke, 1990).

Population within one mile of an observation is an implicit measure of population density and has a negative impact on sales price. Other factors significantly reducing the sales price are the percentage of the population that is under age 18 and our measure of income dispersion in the neighbourhood (squaring the difference between median income and average income). The sales price of a home, *ceteris paribus*, is higher in neighbourhoods with more white and Hispanic residents and those with more white-collar workers, a result consistent with Crane's (1991) study of epidemic theory in a neighbourhood context. The second column shows the high-crime area interaction variables for house and neighbourhood variables. For example, column one shows the square foot coefficient is 0.0005 and that in high-crime areas this coefficient rises by an insignificant 0.000015.

The cost of crime coefficients for the entire sample have the expected negative sign and are modest in magnitude, with the cost of violent crime being significant at the 0.10 level and property crime significant at the 0.01 level.[2] Most striking is the fact that the coefficients reveal a very

Table 11.2   Hedonic estimates of the willingness to pay for public safety
(dependent variable: log selling price of the home)

Independent variables	Weighted crime index hedonic model coefficient estimates	High-crime area interaction coefficient
Intercept	10.59*	
Property-specific variables		
Square footage	0.0005*	0.1566E-4
Lot size	0.0335*	0.1695*
Bedrooms	−0.0238**	0.0107
Bathrooms	0.0246***	0.0462*
Log (age of the home)	−0.0643*	−0.0211
Pool	0.0542*	−0.0174
Covered parking spaces	0.0920*	0.0109***
Fenced property	0.0137*	0.0703**
Fireplace	0.0618*	−0.0095
Central heating	0.1245*	0.0241
Central air conditioning	0.1192*	−0.0333
Vacant	−0.0912*	0.0533***
Assumable mortgage	0.0136	0.0209
Waterfront property	0.1542*	0.0104
Gated community	0.1690*	−0.1486***
Neighbourhood variables		
Population	−0.1197E-4*	0.8497E-5
Percentage white	0.2957*	−0.0979
Percentage Hispanic	3.6638*	0.6278
Percentage bachelors degree	0.0791	−1.0494***
Percentage owner occupied	−0.1064	0.4255***
Percentage kids (17 and under)	−1.3653*	2.2850***
Percentage young adult (18–24)	−1.0836*	2.9819*
Percentage older (55 and over)	−0.0527	2.7457*
Median household income ($)	0.3908E-5	0.2005E-5
(Median-average income)^2 ($)	−.75615E-12*	−2.0003E-10*
Population growth (1990–96)	−0.3941	0.3276
Percentage white-collar	0.4786**	2.0079*
Crime variables		
Log(cost of violent crimes in $)	−0.0163***	
Log(cost of property crimes in $)	−0.0289*	
High-crime area	−3.0699*	
Adjusted R squared		0.83

*Significant at 0.01 level (two-tail).
**Significant at 0.05 level.
***Significant at 0.10 level.

234 The Economic Dimensions of Crime

modest impact on sales price. A 10 per cent decrease in the cost of property crime raises the sales price of a home by $275 and a similar decline in violent crime increases the predicted sales price by $155. The relatively small cost of violent crime coefficient may be due to the fact that victims know the offenders in about half of violent crimes in the US. In these cases the location of residence may be less salient, thereby at least partly accounting for this result.

The finance literature shows that the annual benefit coming from a permanent change in property values is equal to the change in property value times the appropriate interest rate. Estimating the annual willingness-to-pay for such an increase in the cost of crime, assuming a 10 per cent interest rate, gives a figure of only $15 and $27 a year for violent crime and property crime respectively. The high-crime area coefficient is both large and statistically significant, suggesting that the average home in the two deciles with the highest cost of crime, *ceteris paribus*, would be discounted about $37 000; from $94 000 to about $57 000. Controlling for high crime areas in Table 11.2 also changes a number of the neighbourhood coefficients.

Homebuyers apparently change their evaluation of some characteristics when the house is located in a high-crime area. A bigger lot, covered parking, and fenced property are each valued more in high-crime areas than elsewhere, probably because each of these attributes represents a barrier between the house and the street. The impact of controlling for high-crime areas is particularly interesting in the case of the gated community variable. In Table 11.2 the gated community coefficient is positive and significant but its coefficient in high-crime areas is negative (–0.149) and is significant at the 0.10 level. The magnitude of this coefficient is only slightly smaller than it is in column one, so the implication is that buyers do not value this attribute in high-crime areas where safety is most needed. Gated communities may offer both security and status to their residents but since these communities contribute less value in high-crime areas than they do on average, it appears that gated communities are valued for their ability to enhance status and perceived marginal increments to safety in low crime areas.[3]

Four neighbourhood characteristics in high-crime areas, percentage white-collar, percentage over age 55, percentage owner-occupied, and the measure of income dispersion are statistically significant with the expected sign. Population stability is an important ingredient in successful neighbourhoods (Bursik and Grasmick, 1993), a factor that is fostered by homeowners and older households and is therefore associated with higher sales prices. More white-collar workers are associated with higher house prices in all neighbourhoods, but this effect is significantly larger in high-crime areas. Similarly, the deleterious effects of income differences in neighborhoods are accentuated in high-crime areas. Three neighbourhood–high-crime interaction coefficients are significant and a sign opposite the expected sign that is shown in column one of Table 11.2. These are percentage with a bachelors degree, percentage with children under 17, and percentage of the population in the 18–24 age bracket, which is most likely to commit crimes.

III Conclusions

Our results suggest that the cost of crime has a trivial effect on the price of the average home sold in the Jacksonville, FL market. Consequently, the average homeowner's implicit price for a 10 per cent decline in violent crime is estimated to be about $15 per year. This low price is markedly different from the actual cost of crime suffered by households in Jacksonville. The total cost of crime per capita in suburban Jacksonville is $311 and if the average household is composed of three persons, the expected cost of crime per home is $933 per year when weighting the frequency of reported crime by the Cohen, *et al.* (1995) cost of crime estimates.[4] Two factors probably account for the disparity between the implicit price estimate and the cost of crime. First, the probability of being victimised is less than one and can be affected by the behaviour of the household. Private investments in protection, and diligence in locking cars and houses, can reduce the probability of being victimised. Secondly, the option of zero risk is simply not available and the implicit price estimate is based on the differences in public safety that can be achieved within the relatively safe areas which are characteristic of most of the housing market.

Although crime does not substantially affect the price of the average home, house values decline dramatically in high-crime areas. Houses in police beats that are in the top two cost of crime deciles are discounted about 39 per cent relative to a comparable house in the other areas. This suggests that there is a discontinuity in the impact of crime on house prices, which is consistent with studies that show contagion and threshold effects in individual decisions that can affect neighborhood quality.[5] Understanding the processes that are responsible for this discontinuity is an important issue for future research.

The results presented here suggest three conclusions. First, weighting crimes by their seriousness significantly improves models of the cost of crime via implicit price models. Second, estimates of crime costs using jury awards may be criticised, but in implicit price models they appear to be superior to alternative measures of crime that do not weight crime by an index of their seriousness. Third, the relationship between social context and house prices needs further investigation, since the results are sensitive to neighbourhood definition and, within the method employed here, the relationship is sensitive to the radius of the neighbourhood chosen since the cost of violent crime coefficient gets larger as the neighbourhood definition expands while the property crime coefficient gets smaller (Lynch, 1998).

Notes

1   The extent to which homebuyers have good information about the safety of neighbourhoods is an important issue for this approach. The implicit price model reveals the extent to which higher crime is associated with lower house prices. Thus, an insignificant or small cost of crime coefficient could be caused by inadequate information about neighbourhood variations in crime, small spatial variations in the distribution of crime, and/or buyer indifference to neighbourhood differences in crime.
2   The sign and significance of the crime coefficient in these models is very sensitive to how this variable is specified. Using the traditional measure of reported crimes the violent crime coefficient has the expected negative sign and is significant, but property crimes are significantly and positively related to house values. This anomalous result is also reported by Case and Mayer (1995) and is almost surely caused by the tendency of residents of more affluent areas to report relatively minor property offences.
3   The actual increment in safety in gated communities might be smaller than expected by the residents of these areas. Gated communities are likely to reduce access to young offenders who respond to readily available targets of opportunity, but they also provide a signal to property criminals that the community is composed of attractive targets. It should be noted that these observations are based on relatively few observations: in the sample there are 82 observations in gated communities, 14 of which are in high-crime areas.
4   Suburban Jacksonville is defined here by the boundaries of the central city before consolidation. Average size of all households in Jacksonville is estimated to be 2.58 in 1995.
5   See, for example, Akerlof (1997), Crane (1991), and Glaeser, Sacredote and Scheinkman (1996).

References

Akerlof, G. (1997) 'Social Distance and Social Decisions', *Econometrica*, 65, pp. 1005–27.
Austin, J. (1996) 'Are Prisons Really a Bargain? The Use of Voodoo Economics' (Washington, DC: National Council on Crime and Delinquency).

Bursik, R. J. Jr and Grasmick, H. G. (1993) *Neighborhoods and Crime: The Dimensions of Effective Community Control* (New York: Lexington Books).

Butterfield, Fox (1996) 'Prison: Where the Money Is', *New York Times*, 2 June, p. 16E.

Case, K. E. and Mayer, C. J. (1995) 'Housing Price Dynamics within a Metropolitan Area', Working Paper no. 95–3, Federal Reserve Bank of Boston.

Cohen, M. A. (1990) 'A Note on the Cost of Crime to Victims', *Urban Studies*, 27, pp. 139–44.

—, Miller, T. R. and Wiersema, B. (1995) *Crime in the United States: Victim Costs and Consequences*, Final Report to the National Institute of Justice, May.

Crane, J. (1991) 'The Epidemic Theory of Ghettos and Neighborhood Effects on Dropping Out and Teenage Childbearing', *American Journal of Sociology*, 96, pp. 1226–59.

Glaeser, E. L., Sacredote, B., and Scheinkman, J. A. (1996) 'Crime and Social Interactions', *Quarterly Journal of Economics*, 111, pp. 507–48.

Hellman, D. A. and Naroff, J. L. (1979) 'The Impact of Crime on Urban Residential Property Values', *Urban Studies*, 16, pp. 105–12.

Lynch, A. K. (1998) 'Towards the Development and Validation of a Monetized Crime Index'. Unpublished PhD dissertation, Florida State University.

—, and Rasmussen, D. W. (1998) 'Is the Cost of Crime Capitalized into Housing Prices?' Florida State University Department of Economics Working Paper.

Moulton, B. R. (1990) 'An Illustration of a Pitfall in Estimating the Effects of Aggregate Variables on Micro Units', *Review of Economics and Statistics*, 72, pp. 334–8.

Rasmussen, D. W. and Zuehlke, T. W. (1990) 'On the Choice of Functional Form for Hedonic Price Functions', *Applied Economics*, 22, pp. 431–8.

Rizzo, M. J. (1979) 'The Cost of Crime to Victims: an Empirical Analysis', *Journal of Legal Studies*, 8, pp. 177–205.

Thaler, R. (1978) 'A Note on the Value of Crime Control: Evidence from the Property Market', *Journal of Urban Economics*, 5, pp. 137–45.

12

Editors' Introduction

We have noted that economics can contribute tools to understand how effectively a system is working. A maddening feature of the criminal justice system is its fragmentation. At the system's margins there are numerous formal and informal organisations whose ambit touches in some way or another on criminal justice. Meanwhile the value-for-money and efficiency-and-effectiveness initiatives which emerged during the 1980s continue to mark the criminal justice environment.

It seems obvious in this context to ask what are the 'real' costs of criminal justice, but the answer tends to have the character of that unhelpful rejoinder, 'how long is a piece of string?' Attempting to 'account for' criminal justice so that the agencies involved can answer fundamental questions such as whether their efforts came to 'profit' or 'loss' in the previous year has led to the criminal justice audit. As Shapland shows, this calls for a methodology combining the approach of economics with disciplines such as operational research, accounting, social administration and criminology.

Shapland notes that such work also helps us to understand the respects in which criminal justice is indeed a 'system'. Efforts to gauge precisely the unit cost of each stage of criminal justice – such things as average costs to provide support to victims during the reporting and investigation stage – have a wider significance. Only by comparing the use of resources with such data can one see the effective priorities of the system, that is, upon what it spends its money. Informed comparison of, say, spending on victim/offender mediation compared with spending on refuges for rape victims, can then be made.

Auditing Criminal Justice

Joanna Shapland

What is a Criminal Justice Audit?

The idea of a criminal justice audit is to draw together data about cases, costs, people and decisions in order to be able to document the use of criminal justice by different people, the activities of all the people and agencies involved in the delivery of criminal justice and the resources being devoted to it. It should provide a snapshot view of criminal justice for a particular area for one period of time, from which it is possible to display the progress of cases through the stages of criminal justice, the key decision points, the costs to different agencies and parties, and the use being made of public criminal justice provision. It should, for example, allow one to see how much, on average, it costs to provide support to victims per victim during the reporting and investigative stage, how much it costs per offender to caution him or her, how much it costs to keep one offender in prison on remand (not per day, but per average remand period). It is only by comparing the use of resources in this way that one can see the effective priorities of our criminal justice system – where it spends its money. It is then possible to compare the amount spent on offenders of different types with the amount spent on victims, or to look at the extent to which each agency contributes to each stage of criminal justice, or to focus on the kind of work which is being done within criminal justice.

We were asked to do this for Milton Keynes, with funding from a consortium organised through Thames Valley Police.[1] Milton Keynes is a rapidly growing town north of London which had a population of around 250 000 at that time. Subsequently, the same methods have been used to look at mediation in neighbour disputes (Dignan, Sorsby and Hibbert, 1996) and are currently being used to look at civil justice in the Sheffield courts (Shapland *et al.*, 1998).

The opportunity for the audit arose because the criminal justice agencies in Milton Keynes, particularly those working with young offenders, had become disenchanted with their ability to respond to crime and criminality, and frustrated by the difficulty of setting up new initiatives. They did not know how, taken across criminal justice as a whole, their resources were being deployed. Each agency could track its own personnel and its own finances, but how did this relate to the work other parts of the system were undertaking? What was the overall impact in relation to offenders and victims? Would it be possible to undertake an audit of the activity of criminal justice in Milton Keynes across all the participating agencies and groups which formed the Youth Crime Strategy Group,[2] as well as relevant other agencies and voluntary bodies?

This was a unique opportunity to look at the whole of criminal justice in one place and how it was working. There has been considerable pressure to improve the connections within the system, to reduce delays and to improve communication within the system and to users. More recently this has culminated in the formation of the Trial Issues Group nationally, with its local counterparts, the move towards case management by courts, fast-tracking of cases involving persistent juvenile offenders, and the proposals in the Crime and Disorder Act 1998 for advice to the police from Crown Prosecutors and for early hearings to speed offenders through court. The new Strategic Plan for the criminal justice system has brought together the Home Office, Lord Chancellor's Department and Attorney General's Department to work towards joint priorities and targets, stressing reducing crime, reducing delay and (though the targets are not yet set for these latter priorities) increasing satisfaction and confidence levels (Criminal Justice System, 1999). However, this focus on speed and on connections does not indicate the weight of activity within criminal justice. What are the different people within criminal justice doing? How much do these various activities cost? This was the purpose of our criminal justice audit.

The tools for such an audit are analyses of financial data, workloads, procedures, performance indicators, management information on outcomes and so forth. Our idea was to bring together all these, rather than seeing them as separate aspects of performance. In the 1980s and 1990s these forms of data have primarily been used in financial management initiatives, with the watchwords being efficiency, cost-cutting and measuring performance. The focus has been on the financial calculus, with performance of public agencies being measured through financial indicators in a cash-limited public sector. However, these financial and performance measures also represent people's hard work to make criminal justice happen. They reveal the priorities of criminal justice, looked at as a whole – not just in terms of money, but in terms of the amount of effort that is being put into the different parts and aims of the system. Because they render the criminal justice system transparent, they allow practitioners and others to see how things are and how they might be changed. We need to distinguish our audit from 'auditing', in the sense of financial auditors' inspection of the accounts of a company, or a criminal justice agency, to determine whether the published accounts faithfully represent the company's/agency's financial transactions. Our audit went far beyond financial data – nor did we inspect the validity of people's financial transactions. We essentially used the products of the financial auditors (the annual accounts of each agency) as our financial data. We should also distinguish our work from 'cost–benefit analysis', which is generally used to evaluate one particular initiative, comparing its costs with its predicted outcomes and benefits, though there are considerable similarities, especially methodological, between our auditing method and the methods generally used in cost–benefit analysis. We were looking at the *routine work* of all the agencies making up criminal justice, rather than at just one initiative. We also did not take the aims of criminal justice as given (or stated by managers) and then compare outcomes against aims, but rather tried to work out inductively what the aims of criminal justice seemed to be, from the activities which its personnel were engaged in. Equally, our audit is not as ambitious as 'economic trend analysis', because it only tries to produce a snapshot at one point in time (or rather, over one period of time, such as one financial year). If, however, the agencies were to be brave enough to repeat the audit at subsequent points in time, then the kind of data which we provide would be ideal as the starting point for such economic analysis.

There are significant difficulties in carrying out such an audit. It requires researchers with both accountancy skills and socio-legal or criminological skills, so we amassed a team which contained people with these different qualifications. Our backgrounds were in accountancy, criminology, law, psychology and sociology. For such a complicated and interlocking process as the criminal justice system it also requires a process of investigation and translation to make the data from the different agencies comparable, so that we could see how the work of the police, for example, relates to that of social services. Above all, it is necessary to set common and agreed boundaries to what we would consider criminal justice: what cases would fall within the audit, the limit of the geographical area, the period of time to be covered and the procedures we would term criminal justice, bearing in mind that the remits of the agencies with which we would be working stretch far beyond criminal justice into civil matters, housing, family support and so on.

Bounding the Audit

Our focus in the Milton Keynes criminal justice audit was on criminal justice as reaction to crime. Full details of the audit can be found in Shapland *et al.* (1995, 1996). We were interested in what might be called the traditional, reactive processes which follow a report of an offence to the police: investigation, relations with victims, the decision to prosecute, diversion of offenders, prosecution, the courts, the penal system. We were hence leaving out of our conception of criminal justice all the preventative and proactive activities of the police and other agencies: those activities which aim to prevent crime or criminality, or to target suspected offenders (surveillance, etc.). We would argue that, given that it is not possible to ignore crimes which have been committed whilst redirecting resources to prevention, it is important to understand how the criminal justice system responds to crime. In fact, the Milton Keynes audit seems to have played some part in redirecting activity away from processing offenders towards more creative work in preventing offending (Home Office, 1997: Introduction).

We took the starting point of the audit as being the point at which a crime report was filled in by the police after the offence had been reported. But which offences and which agencies to include? Establishing a common basis for the offences we would include was very difficult. The kinds of offences included in the *Criminal Statistics* compiled by the police (for example, the list of 'notifiable offences') are different from the list of offences included in the reports by courts of defendants who are convicted and given various disposals (which are generally the 'indictable and triable either way offences',[3] though summary offence results are collected and provided in supplementary tables) (see Appendices 3 and 4 in Home Office, 1998). We used 'notifiable offences' as the offences we would include in the audit. We hence had manually to adjust all the figures from the prosecution and court accordingly. Until definitions are aligned, this will be a problem for all those doing cost–benefit analysis or economic analysis from national or court statistics.

The major emphasis for Milton Keynes was on youth offending, so we needed to separate out outcomes, processes and resources devoted to different age groups. For the audit we used the age ranges 10–13, 14–17 and 18–25, as well as calculating the figures for all offenders, irrespective of age.

We found it extremely difficult to set fast rules as to which agencies to include and which to omit. Some are central criminal justice agencies, as far as reactive activity is concerned. These included Thames Valley Police, the Crown Prosecution Service (CPS), Milton Keynes Magistrates' Court, the Lord Chancellor's Department,[4] Buckinghamshire Probation Service, Victim Support, the Home Office (attendance centres[5]), the Prison Service and the social services and education departments of Buckinghamshire County Council. There were, however, a number of voluntary and statutory agencies which provided facilities for offenders, either diverted from the courts, or on court orders, such as the Milton Keynes Young Befriender Scheme, which we also included, but which are likely to be run by different agencies in different towns. However, we could just as well have argued for the inclusion of a number of other agencies which, for practical reasons,[6] we did not include. These were the various parts of the health service, the High Court, Women's Aid, Rape Crisis, the Women's Royal Voluntary Service (court refreshments), the Ark (accommodation for young people), accommodation provided by the housing department of the council, the Salvation Army and other voluntary organisations, and the Samaritans. Again, these will be decisions which will be faced by all other researchers seeking to evaluate the scope of, or initiatives taken by, criminal justice, particularly as such initiatives become more likely to be taken on an inter-agency basis.

One of the greatest difficulties in researching criminal justice is the lack of comparability of the geographical areas over which different agencies work. We were fortunate in that, because Milton Keynes is a relatively new city, built largely on a greenfield site, many agencies had the same boundaries, which tended to be the Milton Keynes petty sessional division (magistrates' court area). This included both the city and some of its rural hinterland. However, the police and CPS, as well as the social services and probation, operated on different areas and so again it was necessary manually to add and subtract villages accordingly.

Finally, we needed to decide which time period we would take. We only had the resources to look at one year's activity and, given the relative inflexibility of the presentation of financial data at that time (only by whole financial year for some agencies), compared to the flexibility of case and offender data (monthly or by date of occurrence), we decided to look at the financial year from 1 April 1993 to 31 March 1994. Luckily all our agencies operated on the same financial year. The audit looks at the work done within that year, rather than tracking cases from that year to their conclusion.

Carrying Out the Audit: Where is the Information?

We have gone into some detail about the decisions made on the boundaries for the audit, because similar decisions will face anyone seeking to carry out audits, economic analyses or cost–benefit evaluations relating to criminal justice or preventing crime. The next set of hurdles was to track down the information we needed on costs, cases and processes.

Increasingly, agencies are able to provide figures for the national average cost of, say, a police officer (Home Office) or a judge (Lord Chancellor's Department). Most economic work on criminal justice has used these national figures (which include employment on-costs and support costs). We would argue, however, that such a national perspective is increasingly inappropriate. Though some agencies will continue to work on a centralist, national basis (most obviously the courts), the logic behind the new youth justice orders contained in the Crime and Disorder Act 1998 and the Youth Justice and Criminal Evidence Bill 1999 is one of local partnerships using schemes which are developed for the local offending population. Those schemes will vary from locality to locality and are highly likely to involve different contributions from statutory criminal justice agencies and voluntary bodies. The cost (including overheads such as premises) of a police officer or probation officer working in such settings is not the same as that for a police officer on patrol or in the CID, or a probation officer working in a prison. The search for evidence-based disposals in criminal justice, including reparative justice, contracts with offenders for community sentences and diversion, will accelerate the trend towards specifically designed programmes for adults as well as juveniles (akin to the 'sentencing packages' which courts have used for some years in the USA). All these programmes and orders need to be costed and evaluated. National figures can provide only a very approximate and biased basis on which to undertake this evaluation.

If, then, one wishes to look at actual costs and personnel numbers, rather than at national estimates, where are these to be found? Our experience in the Milton Keynes audit was that though they can relatively easily be acquired, the data are not currently being held in such a way that facilitates these calculations. Financial data for statutory criminal justice agencies are largely held on systems which use the standard local authority financial codes. Each functional unit is likely to be a cost centre (for example, the police dog section, or traffic, or the community service unit in probation), which facilitates matching up spend with personnel and their functions. However, support services and premises are all costed separately. To calculate the full cost of, for example, the provision of community service, it is hence necessary to work out all the support services used across the probation service and apportion these costs (which we have called 'indirect costs'). Premises costs, as well as running costs (travel, photocopiers, telephone systems, etc.) vary considerably between buildings, depending upon the leasing arrangements which the agency has entered into. Premises and running costs have to be apportioned between workers in those premises. At the time we were doing the audit no criminal justice agency had brought together their salary and their premises/running costs so that they had an idea of what it cost to have a particular person at a particular grade in a particular building. In many agencies, spend on salaries was being tied up only with spend on buildings, running costs, maintenance etc. at the level of the entire agency, or geographical sector.

The reason why spend on different elements may not be matched up lower down in the process lies in the increasing centralisation of purchasing, fleet management and buildings management in criminal justice agencies. As these are seen more and more as specialist functions, so their budgets become held and managed centrally, so that spend can be compared against budget for that function. The downside is that agencies are not able to work out speedily what it would cost to take new, local initiatives, or to evaluate the initiatives they have taken in cost–benefit terms. The trend towards evidence-based social policy is in tension with the trend towards professional management of service functions within criminal justice agencies. As a result the criminal justice system has become not only disjointed in terms of comparisons between agencies, but fractured in terms of comparisons between different parts of the same agency. For any effective move towards evaluation of local initiatives to take place, we think these deficiencies in tying up spend figures need to be remedied. This means that, within every agency, it needs to be easy to equate resource costs with activities at both agency level and at local budget-holder management level. It is obviously also vital that the same conventions are used across the criminal justice sector in relation to indirect costs.

For the audit we had to take the costing one stage further – to match up the work people were doing with its cost, so that we could calculate the unit cost of each stage of criminal justice. Creating a unit cost per offence, offender or victim meant matching up the numbers of offences, offenders and victims involved at each stage of criminal justice with the work of the professionals and volunteers for each stage and their cost. This meant acquiring the figures from all the case processing systems for each agency, from the command and control, crime statistics, and custody computer systems of the police, through the CPS and court management systems, those of the Probation and Prison Services, to the figures given in the annual reports of the voluntary agencies for the same time period. In some instances we were able to obtain figures only by interviewing relevant staff and getting them to look at their unit's log book or record cards (for the police underwater search team, for example). In others we had to undertake a manual analysis of a sample of files (to work out the average number of different kinds of cases in a magistrates' court morning, for instance). Our detailed methods and the assumptions we had to make where records were not available at a sufficient level of detail are set out for each part of the criminal justice system in the full report of the audit (Shapland *et al.* 1996).

What Criminal Justice Is: Discretionary Decision Making

Once we had these detailed calculations we could start adding up criminal justice in Milton Keynes – by agency and by stage of the process. The total was surprisingly large: over £16 million for 1993/94, without including child abuse/family protection cases.[7] This related to 27 005 notifiable offences, giving an average cost per offence of £595. However, as is well known, the attrition diagram for cases as they go through the criminal justice process is a very rapidly thinning 'tree-trunk': of our 27 005 offences which were recorded by the police as notifiable offences in that year, just 20 per cent were 'cleared up' (i.e. the offender was detected). The 5410 offences cleared up were put down to 4708 offenders who were arrested, giving an almost 1:1 relation of the number of offenders to individually recorded offences (though obviously one offender could be identified for more than one separate offence during the time period). However, just 2563 offenders were identified at the CPS stage (equivalent to 9 per cent of notifiable offences), with the rest being released with no charge (1068) or given a caution (1057). A few had their prosecutions dropped or were otherwise dealt with before trial, so that 1841 notifiable offenders were tried at court (7 per cent of notifiable offences), leading to 6 per cent being found guilty at the magistrates' court (1667 offenders) and 0.9 per cent being sent to the Crown Court, with 181 being found guilty at the Crown Court.

The cost of criminal justice is very unevenly distributed between the stages of criminal justice. Overall, just 5.89 per cent of the £16 million was spent on recording crime and assisting victims (including police and Victim Support work on assisting victims). The largest slice of cost was involved in investigating crime and deciding whether to prosecute (46.82 per cent), the bulk of it by the police (over £7 million), with much smaller amounts by the CPS, social services, education and diversion involving the voluntary sector. The work of the courts and those involved at court (police, witness service, CPS, social services, the Probation Service and Prison Service) accounted for 28.39 per cent of total cost, and post-court disposals (prison, community sentences, financial penalties and their administration, conditional release, etc.) 18.90 per cent. In relation to juveniles the major elements were the cost of dealing with suspected child abuse (over £2 million) and secure accommodation for the very few young people who were remanded there or required residential accommodation during their sentences (nearly £700 000), whilst few resources were devoted to community programmes for juvenile offenders (just over £100 000) or to pre-sentence inter-agency liaison on juveniles (under £200 000).

However, because the numbers of offences, offenders and victims vary so much between stages, the average unit cost of each stage does not follow the overall pattern of cost. So, for example, it cost just £20 per offence to record a notifiable offence and a mere £5 to assist a victim. The cost of investigating a notifiable offence was more substantial at some £152 per offence, with that of arresting an offender £177 per offender, preparing the prosecution file £126, inter-agency liaison for juveniles £105 per juvenile, administering a caution £36, and the CPS reviewing the case £51 per offender. The whole court process was far more expensive, at some £2 094 per offender (including the cost of all remands – though obviously this was largely driven by the cost of remands in custody for those few offenders who were so remanded). The post-court process costs varied enormously by disposal, though every conviction entailed a £21 cost per offender to record the conviction. So discharges, fines and suspended sentences (ignoring any subsequent offending) effectively had zero cost (the financial benefit of fines was equivalent to the cost of their administration), but community sentences averaged out at £4002 per offender and custody at £8037.

The Impact on Victims and Offenders

Expenditure figures give an idea of what the priorities of the system are, but it is necessary to look at the kinds and amounts of contact that victims and offenders have with criminal justice to see justice from their viewpoint. If we think, first, of victims' contacts with the system, it became clear from the results of the audit that victim contacts are almost entirely concentrated in the early stages of the system (less than 5 per cent of victims' offences led to a court appearance). Victims' contacts were through the initial recording of offences, which might be just a matter of a telephone call with the police Crime Desk and, for a few victims (primarily victims of burglary and of serious violent attacks) contact with Victim Support, with police Area Beat Officers or with Crime Prevention Officers. Victims' contact with criminal justice was both minimal and extremely frugal in terms of expenditure.

What was surprising to us, however, was that, although offenders were clearly much more costly to criminal justice, their amount of contact with people from the criminal justice system was also often minimal. What contact there was generally involved people obtaining information from offenders (police during the investigation, solicitors, probation, social services, courts, inter-agency liaison, etc.), rather than talking about the offending behaviour. It was not unless offenders obtained one of the more serious and onerous penal measures that time might be spent addressing their offending (on a community sentence, or in custody, or in social services accommodation). By that stage, of course, offenders would normally have several previous convictions.

Most of the activity of criminal justice, in the sense of the reactive response to offending, was in essence processing. People employed in criminal justice spent their time trying effectively and efficiently to process the case which they had received, amassing information in order to pass the case on to the next stage in the process. Unfortunately, very little of this effort was devoted to the offender or the victim, as opposed to the offence. The system was prioritising decision making about the offender, rather than the offender himself or herself. Consultation was between different agencies in order to take the relevant decision with the best information available – and according to due process safeguards.

The major question for the Milton Keynes agencies, as it is for the criminal justice system generally, was whether this is the right priority. Of the £16 million, should less than £200 000 be spent on victims (including providing crime prevention advice and all support) and a maximum of just under £3 million on offenders (including both diversion measures and all penal measures, and counting the whole cost of prison as addressing offending behaviour)? We concluded in the report to the Milton Keynes Youth Crime Strategy Group that the system was primarily occupied with speeding the 'passing parade' of offenders and cases. It did not seem to be intending to have a direct impact, through contact, on offenders. If the criminal justice system was aiming at reducing crime, it was clearly trying to do so only indirectly.

The Milton Keynes agencies concluded that there was a need for a major change in priorities, towards crime reduction through working with offenders and potential offenders, and through detection. The audit led to a programme of work within the city. Nationally, however, we still have to question the priorities of criminal justice spending. It is of course important to react to offending – we cannot abolish criminal justice. But we also have to develop a system which aims to work directly with offenders and with victims. A contactless criminal justice system can never hope to deliver reductions in crime. In such a context the investment in central initiatives such as the Home Office Reducing Crime Initiative of £250 million over three years still seems a relatively small sum to put into work to reduce offending. It is equivalent to just £407 223 for Milton Keynes (reduced proportionately by number of notifiable offences per year). In terms of work with victims, the increases have been even smaller. We still have primarily an inwards-facing system, one that prefers to talk to itself, rather than a system which faces outwards to the community and talks to both offenders and victims.

Notes

1 The consortium included Marks and Spencer PLC, the Home Office, Buckinghamshire County Council, Milton Keynes Borough Council, Safeway PLC, the Commission for New Towns and Mercedes Benz.
2 The Youth Crime Strategy Group consisted at that time of representatives from Thames Valley Police, Social Services, the Probation Service, the Crown Prosecution Service, Milton Keynes Borough Council, the magistrates, the Clerk to the Magistrates and the Education Department of Buckinghamshire County Council (from the education management side, from the Educational Welfare Service and from Youth and Community).
3 Indictable offences are those which can be tried only at the Crown Court (the most serious offences, such as murder, rape and robbery). Triable either way offences may be tried at the Crown Court or the Magistrates' Court.
4 At the time the Lord Chancellor's Department was directly responsible for the Crown Court centres to which cases were sent from Milton Keynes, primarily Aylesbury and Northampton. Today, the Court Service would be the relevant agency.
5. Young offenders may be sent to an attendance centre for a period of hours every week for several months, normally on a Saturday afternoon, to undertake a programme designed to deter offending and address offending behaviour. Centres are often run by the police.
6 The major problem was the great difficulty of disentangling services for offenders from other services provided by these agencies.
7 If child abuse/family protection cases are included the total rises to nearly £19 million. The difficulty of including these cases is that much of the expenditure is incurred during the investigation, when it is not clear whether a notifiable offence has occurred. The social services did not record how many such cases they investigated in 1993/94, but the police files indicated that 864 victims were involved. The direct costs (not including support and premises) for the police were £303 320 and for social services were £2 224 444. Police worked in a joint team with social services on child protection matters.

References

Criminal Justice System (1999) *Strategic Plan, 1999–2002* (London: Lord Chancellor's Department, Home Office and Attorney General's Department, March; also http://www.criminal-justice-system.gov.uk).

Dignan, J., Sorsby, A. and Hibbert, J. (1996) *Neighbour Disputes: Comparing the Cost-Effectiveness of Mediation and Alternative Approaches* (Sheffield: Centre for Criminological and Legal Research).

Home Office (1997) *No More Excuses – A New Approach to Tackling Youth Crime in England and Wales*, White Paper (London: Home Office, Cm 3809; also http://www.homeoffice.gov.uk/nme.htm).

Home Office (1998) *Criminal Statistics England and Wales 1997* (London: HMSO, Cm 4162, 1998).

Shapland, J., Hibbert, J., l'Anson, J., Sorsby, A. and Wild, R. (1995) *Milton Keynes Criminal Justice Audit: Summary and Implications* (Sheffield: Institute for the Study of the Legal Profession on behalf of the Milton Keynes Youth Crime Strategy Group).

Shapland, J., Hibbert, J., l'Anson, J., Sorsby, A. and Wild, R. (1996) *Milton Keynes Criminal Justice Audit: The Detailed Report* (Sheffield: Institute for the Study of the Legal Profession).

Shapland, J., Sorsby, A. and Hibbert, J. (1998) 'Towards a Civil Justice Audit', Report to the Lord Chancellor's Department, unpublished document (Sheffield: Institute for the Study of the Legal Profession).

Annotated Further Readings

Box, S. (1987) *Recession, Crime and Punishment* (London: Macmillan).
Integrates major theories in criminology to explain causal links between economic factors and crime. Reviews empirical studies in North America and Britain using aggregated data to explore relationship between unemployment and crime and income inequality and crime. Identifies key methodological problems encountered when using time-series, cross-sectional and longitudinal research designs.

Chiricos, T.G. (1987) 'Rates of Crime and Unemployment: an Analysis of Aggregate Research Evidence', *Social Problems*, 34, pp. 187–212.
Reviewing evidence from 63 empirical studies of the relationship between unemployment and crime, challenges the findings of earlier reviews which created a 'consensus of doubt' by describing the relationship as insignificant and inconsistent. Explores the conditional nature of the unemployment–crime relationship and shows that studies undertaken since 1970 reveal a positive and frequently significant relationship between unemployment and crime, particularly in the case of property crime.

Field, S. (1999) 'Trends in Crime Re-visited', *Home Office Research Study 195* (London: HMSO).

Freeman, R. (1983) 'Crime and Unemployment', in J. Q. Wilson (ed.), *Crime and Public Policy* (San Francisco, CA: ICS Press).
Finds measures of deterrence and crime more strongly linked than unemployment and crime, with relationship stronger for property than violent crime.

Hagan, J. (1993) 'The Social Embeddedness of Crime and Unemployment', *Criminology*, 31(4), pp. 465–91.
Deals with the fact that while many macro-level empirical studies reveal that unemployment leads to crime, individual-level data from ethnographic studies suggest the causal sequence can be in the opposite direction. Using ethnographic research and a panel study using individual-level data, Hagan suggests the concept of 'criminal embeddedness' to explain the causal chain whereby youths become isolated from lawful employment networks.

Hale, C. and Sabbagh, D. (1991) 'Testing the Relationship between Unemployment and Crime', *Journal of Research in Crime and Delinquency*, 28(4), 400–29.
Changes in unemployment are positively associated with changes in crime while changes in unemployment a year earlier have no effect.

Symposia (1996), 'The Economics of Crime', *Journal of Economic Perspectives*, 10(1).

Thornberry, T. and Christenson, P. (1984) 'Unemployment and Criminal Involvement: an Investigation of Reciprocal Causal Structures', *American Sociological Review*, 49, pp. 398–411.

Asserts many explanations of criminal aetiology are flawed by being unidirectional and overlooking the theory that criminal involvement can be the result of reciprocal causal influences. Adopting a linear panel model approach to the study of unemployment and crime, they argue criminal behaviour is not simply the product of unemployment, but unemployment and crime mutually influence one another over the person's lifetime.

Viscusi, W. K. (1986) 'Market Incentives for Criminal Behaviour', in R. Freeman and H. Holzer (eds), *The Black Youth Employment Crisis* (Chicago, IL: University of Chicago Press).

Belief that rewards from crime exceed those from lawful work was strongly associated with criminal involvement. Includes actual annual income figures from crime.

Witte, A. (1997), 'The Social Benefits of Education: Crime', in Jere Behrman and Nevzer Stacey (eds), *The Social Benefits of Education* (Ann Arbor, MI: University of Michigan Press), pp. 219–46.

Index

Abrahamse, A., 170, 172
Actus reus, 84
aetiology, aetiological, 2, 9, 14, 121, 141, 206, 217, 219
age structure, 154, 235, 242
aggregate level studies, 218–20, 251
Akerlof, G., 188–9, 236
alcohol, 77, 80
Allan, E., 94, 97, 213, 220
Andvig, J., 119
anomie, 16, 22
Arrow, K., 45, 67
assault, 16, 17, 31, 70, 73, 76–7, 144, 180, 195, 203, 230
audit, 8, 238–43, 245–8
Austin, J., 227, 236

Baker, J., 188, 190
Batani-Khalfani, A., 167, 175
Beattie, A., 215, 222
Beccaria, Cesare, 51, 60
Beck, A., 152, 155, 156, 173, 174
Becker, G., x, 1–3, 5, 9, 13–15, 61–2, 66, 67, 69, 71–5, 78, 80, 83, 87, 96, 97, 99, 122, 136, 142, 149, 177, 186, 188, 189, 193, 207, 218, 220
Bell, D., 109, 111, 119
Bentham, Jeremy, 50, 51, 59, 62, 63, 67
Berk, R., 179, 191
Blau, J., 211, 220
Blau, P., 211, 220
Block, A., 109, 119
Block, M., 87, 97
Blossfeld, H.-P., 195, 207
Blumstein, A., 74, 80, 186, 188, 189
Boggess, S., 153, 172, 173
Bonger, Wilhelm, 144, 147, 210, 220
Borooah, V., 213, 220
Bound, J., 153, 172, 173
Box, S., 93, 97, 142, 145, 147, 213, 220, 251
Brenner, H., 212, 220
Brier, S., 188, 189

British Crime Survey, 92, 218
'broken windows hypothesis', 8
Buchanan, C., 71, 72, 80
burglary, 7–8, 15, 19, 31–2, 46, 73, 75, 77, 83, 87, 92–6, 133, 143, 157, 159, 161, 170, 211–13, 217, 229–30, 248
Bursik, R., 234, 237
Butterfield, F., 227, 237

Cagan, P., 59, 67
Cameron, S., 73, 80, 92, 97
Cantor, D., 93, 97, 159, 174, 212, 219, 220
Cappel, C., 159, 173
Carr-Hill, R., 73, 80, 92, 93, 97, 212, 220
cartel, 3, 99, 101–5, 107–14, 116–18
Case, K., 236, 237
Chamard, S., x, 4, 121, 122
Chamberlain, G., 183, 189
Chin, K.-L., 167, 173
Chiricos, T., 93, 97, 157, 160, 161, 172, 173, 212, 220, 251
choice
    allocative, 1, 3, 13, 225
    rational, see rational choice
Christenson, R., 163, 168, 179, 180, 191, 214, 222, 251
Chubb, J., 115, 119
civil law, 69, 70, 239
Clark, A., 93, 97
Clark, K., x, 4, 121, 122
Clarke, A., x, 1, 93, 98, 210, 217, 222
Clarke, R., 122, 128, 129, 136, 137
Clarkson, C., 84, 97
Clear, T., x, 8, 225, 226
clear-up rate, 3, 20, 74, 246
Cohen, L., 161, 174, 189, 211, 212, 220, 221
Cohen, M., 227, 228, 229, 230, 235, 237
cohort study, 6, 176, 177
Collins, J., 187, 189
Collins, O., 213, 220

consumption, 76, 145–6
  personal, 5, 141–4, 147, 214
contagion and threshold effects, 8, 225, 235
contingent valuation method, 227
Cook, P., 75, 80, 181, 188, 189
Cornish, D., 122, 137
cost, marginal, 8, 26–7, 226
costs and benefits, 3, 13, 22, 71–2, 78, 84, 121, 124, 133, 218, 225, 227, 241–2, 244–5
Cox, L., 77
Crane, J., 232, 236, 237
Cremer-Schafer, H., 202, 203, 207
crime as work, crime-as-work models, 6, 176, 179, 188
crime
  earnings from *see* returns to crime
  organised, 3, 4, 99, 120–1, 135: *see also* syndicates, syndicated crime
  personal, 5, 17, 141, 144, 146
  prevention, 1, 4, 8, 42, 72, 121–3, 126, 128, 130, 133, 135–6, 242; situational, 121–2, 124, 126, 128–9, 132
  property, 3, 5, 17, 69, 71, 73–8, 82–3, 87, 93–4, 96, 141–7, 150, 161–2, 180, 195, 211, 213–14, 218, 228–9, 231–4, 236, 251
  returns to, returns from *see* returns to crime
  sexual , 74, 93, 144, 146
  strict liability, 70
  violent, 3, 8, 69, 70, 72–4, 76–7, 93, 114, 144, 146, 150, 193, 195, 197, 210–11, 213, 225–8, 231–36, 248, 251
criminal
  career, 9
  opportunities, *see* opportunities, illegitimate
  record, 5, 168, 185, 187, 205
Criminal Justice System, 240, 250
criminology, 1–4, 8, 14, 16, 51, 63, 84, 99, 121–2, 130, 136, 214, 218, 238, 241, 251
cross-section, cross-sectional data, 7, 76, 92, 96, 149, 159, 160–1, 163, 179, 194–5, 212–13, 215, 251

Crowley, J., 180, 189
Curtis, L., 211, 220

Dale, M., 168, 173
Danziger, S., 211, 220
Dau-Schmidt, K., 90, 97
Deadman, D., 76, 79, 80–1, 92–3, 96–8, 213, 222
death penalty, *see* punishment, capital
decision-making, 2, 4, 13, 121, 123, 127, 133, 141, 246, 248
demand, 2, 97, 112, 128, 131–2, 147, 176, 188
  curve, 47, 127, 134–5
desistance, 209
deterrence, 3, 13, 16, 25, 40, 49, 69–73, 75, 77–9, 84, 90, 92, 96, 121, 133, 150, 154, 178–80, 182, 184, 187, 193, 218–9, 251
  hypothesis, 3, 5, 69–75, 77, 82
Devine, J., 214, 219, 221
Dickens, W., 188, 189
Dickinson, D., 213, 221
Dietz, G.-U., 199, 208
differential association, 22
Dignan, J., 239, 250
diminishing returns, 19
disadvantage, 7, 201, 210–12, 214–15
disaggregation, 7
displacement, 73
Dnes, A., x, 3, 5, 69, 70, 73, 80, 82, 85–6, 99
domestic role, 7
drugs, drug trade, drug trafficking, 17, 135, 149–50, 153, 156, 170, 180, 188, 195
Dwayne-Smith, M., 214, 221
dynamic model, *see* model

earnings, 5, 6, 92, 218
  see also wages
econometric, 217
educational achievement, 6, 76, 176
Ehrlich, I., 23, 28, 30, 46, 60, 67, 74, 77–8, 80, 92, 97, 177, 189, 193, 208, 218, 221
elastic, elasticity, 2, 3, 27–8, 32–3, 74–5, 135–6
Elder, G., 173

Erez, E., 188, 189
ethnic, 151, 166
  see also race
ethnography, 99, 149, 166–7, 172, 178, 251
event history analysis, 192, 195
expected utility, *see* utility, expected
expenditure, consumer, 17
extortion, 100–1, 104, 109–10, 116

Fagan, J., 166, 173, 194, 208, 217, 221
Farrell, G., x, 4, 122
Farrington, D., 9, 146, 148, 164, 173, 179, 186, 190, 219, 221
felony, 16–17, 21, 23, 28–32, 39, 46, 50
Felson, M.. 186, 190, 212, 220, 221
female, women, 6, 192, 196, 200
Ferguson, R., 168, 173
Ferri, Enrico, 210, 221
Field, S., x, 5, 76, 77, 81, 93, 96, 97, 122, 137, 141–2, 148, 149, 214, 221
Fielding, N., x, 1, 93, 98, 210, 217, 222
Fienberg, S., 188, 189
Figlio, R., 163
fines, 16, 22, 24, 34–42, 46–7, 49, 50, 63–4,78, 247
Finn, R., 168, 173
Fiorentini, G., 1, 10, 100, 101, 119
firm, theory of, *see* theory of the firm
Flinn, C., 186, 188, 190
Fontaine, P., 168, 173
formal model, 2, 13, 82, 86, 136, 141, 225
  see also mathematical model
Fowles, R., 211, 221
Fox, J., 212, 221
Franchetti, L., 105, 106, 119
fraud, 17, 83, 117, 119
free rider, 106
Freeman, R., x, 5, 6, 93, 98, 149, 150, 151, 157, 161, 163, 165, 168–9, 172, 173, 176, 188–90, 194, 208, 212, 214, 221, 251
Friedman, A., 111, 119
Friedman, J., 101, 119

Galesi, Y., 167, 175
Gallagher, B., 148, 173, 190, 221

Gambetta, D., x, 3, 4, 10, 99–102, 104–5, 107–9, 114, 119, 121
gender, 6, 195, 198, 200, 206
General Household Survey, 92, 215
Gilliard, D., 152, 155–6, 174
Gimbel, C., 173
Glaeser, E., 236, 237
Good, D., 165, 174, 179, 188, 190
Gosch, M., 111, 119
Gottfredson, D., 179, 186, 190
Grasmick, H., 234, 237
Greenwood, P., 180, 191
Griesinger, H., 163, 175, 182–3, 185, 191
Grogger, J., 165, 169, 174
gross domestic product, 82, 93, 95–6

Hagan, J., 168, 174, 214, 221, 251
Hakim, C., 93, 98
Hale, C., 93, 96, 97, 158, 174, 213, 221, 251
Hammer, R., 111, 119
harm, concept of, 18, 25, 40–2, 70–1, 84–5
Harper, F., 64, 65, 67
Hartley, P., 71, 72, 80
Hawkins, G., 172, 175
Heckman, J., 189, 190
hedonic model, 228–30
Heineke, J., 87, 97, 188, 190
Heinz, W., 203, 208
Hellman, D., 228, 237
Henry, A., 96, 98
heroin, 105
Herrnstein, R., 72, 81, 189, 191
Hibbert, J., 239, 250
Hill, R., 211, 221
Hindelang, M., 163, 174
Hirschi, T., 163, 174, 180, 190
Holland, P., 195, 208
Holzer, H., 165, 167, 174
Holzman, H., 181, 190
Homel, R., 122, 128, 129, 136
Home Office, 242, 250
homicide, *see* murder
Hope, T., 122, 137
Howsen, R., 172, 174

illegitimate opportunities, *see* opportunities, illegitimate

implicit price model, 228, 235, 236
imprisonment, 16, 18, 22, 24–5, 33,
    40–3, 46, 49, 50, 63, 77–8, 151–2,
    154–6, 169, 171–2, 200, 203, 225,
    239, 247–8
incarceration, *see* imprisonment
indicator
    coincident, 146
    lagging, 146
indifference curve, 88, 90
inelastic, 2, 111, 112
inequality, 142, 145, 149–50, 156,
    159–63, 169, 200, 209–12, 215,
    217–20, 251
informal economy, 4
insurance, 7, 71, 226
Ivie, R., 173

jail, *see* imprisonment
James, F., 64, 65, 67
Jankowski, M., 167, 172, 174
Jarrell, S., 172, 174
job market, *see* labour market
Johnson, T., 66, 67
Jupp, V., 219, 221
juvenile, 24–5, 34, 183, 185, 187, 193,
    203–4, 207, 240, 247
    see also youth

Karni, E., 190
Kelling, G., 8, 10, 131, 137
Kersten, J., 199, 208
Kessler, D., 186, 190
Kleinman, E., 61, 67
Krivo, L., 211, 221
Kronauer, M., 200, 208

ℓ'Anson, J., 250
labour market, 2, 4–6, 9, 79, 96, 139,
    141, 145, 149–50, 156–7, 159–60,
    163–4, 166–72, 186, 192, 197, 198,
    202, 206, 213–14, 218
labour market discipline, *see* work
    ethic
labour supply, 3, 82, 87, 123, 165
laissez faire, 15
Land, K., 93, 97, 159, 161, 174, 211,
    212, 219, 220, 221

Landes, W., 66, 67
Landesco, J., 109, 120
Langan, P., 154, 174
Lattimore, P., 188, 190
Laub, J., 168, 169, 175, 196, 208
Lavin, M., 180, 191
Leamer, E., 79, 81
Lee, D., 160, 161, 162, 165, 174
legitimate opportunities, *see*
    opportunities, legitimate
Lemert, E., 9, 10
Lempert, R., 79, 81
Lenihan, K., 179, 191
Levi, M., 4, 10
Lewis, D., 92, 98
Lindenberg, S., 193, 208
Loftin, C., 211, 221
Long, S., 93, 98, 142, 148, 166, 174–5,
    188–91, 212, 221
longitudinal, 6, 69, 92, 149, 157, 168,
    169, 192, 194, 251
Lynch, A., x, 8, 225–6, 228–9, 236–7

MacCoun, R., 170, 175
MaCurdy, T., 189, 190
mafia, mafiosi, 4, 99–107, 109–14,
    116–19
Maitland, F., 63, 68
male, men, 5, 6, 9, 147, 151, 154–5,
    157–8, 163–4, 167–8, 170–2,
    176–8, 181, 187, 192, 196, 199,
    201–3, 213
manslaughter, *see* murder
marginal return, *see* return, marginal
marginal utility, *see* utility, marginal
marginal cost, *see* cost, marginal
Marshall, A., 59, 68
Martin, S., 189
mathematical model, 2, 13
    see also formal model
Matt, E., 208
Maung, N., 218, 221
Mayer, C., 236, 237
Mayhew, P., 218, 221
maximiser
    rational, 69, 71
    utility, 78, 123
maximising behaviour, 70
McCall, P., 161, 174, 211, 221

measurement, 3, 5, 141, 182, 240
  error, 92, 97, 162, 218
mediation, victim/offender, 238–9
mens rea, 84
Merva, M., 211, 221
Messner, S., 174, 211, 221, 222
method, methodology, 4, 209–10, 238,
  241, 251
Miller, T., 227, 237
Mirlees-Black, C., 218, 221
model 122, 141, 176, 178, 183, 186,
  189, 209, 218, 223, 226, 228, 232
  dynamic, 6
Moene, K., 119
Montmarquette, C., 179, 180, 190
Moore, J., 172, 174
Morley, L., 148, 173, 190, 221
Moulton, B., 230, 237
multiple regression, *see* regression,
  multiple
murder, 15–17, 19, 21, 29, 31, 33, 39,
  46, 49, 70, 73, 76–9, 84, 86, 153,
  161, 211–12, 229–30, 232, 249
Murphy, P., 170, 175
Murray, C., 77, 81
Myers, S., 179, 180, 181

Nagin, D., 73, 74, 80, 81, 189
Naroff, J., 228, 237
National Crime Victimization Survey,
  153, 226, 228
Needels, K., 169, 175
Nerlove, M., 179, 180, 190
New Jersey State Commission of
  Investigation, 110, 120
New York Organised Crime Task Force,
  110, 112, 113, 120
norm, normative, 15

opportunities
  illegitimate, 6, 124, 127–8, 130–1,
    135–6, 156
  legitimate, 6, 156, 218
opportunity
  costs, 75, 78, 161, 171, 194
  locus, 88, 89
optimality, optimality analysis, 19, 21,
  25–6, 28, 33, 50
organised crime, *see* crime, organised

Orme, J., 222
Orsagh, T., 142, 148
Osborn, D., 96, 98
Osterland, M., 193, 198, 208
Oswald, A., 93, 97

Padilla, F., 166, 175
Palloni, A., 168, 174
panel study, 176–7, 179, 181, 192, 195,
  212, 219, 251
Parker, R., 211, 221
Passell, P., 77, 79, 81
Patterson, E., 211, 222
Pease, K., x, 4, 121, 122, 133, 137
Peltzman, S., 1, 10, 100, 118, 120
personal consumption, *see*
  consumption, personal
Petersilia, J., 150, 180, 191
Peterson, R., 211, 221
Phillimore, P., 215, 222
Phillips, L., 181, 191
Phlips, L., 186, 189, 191
Pirog-Good, M., 165, 174, 179, 188,
  190
Plant, A., 45, 68
Pollock, F., 63, 68
Pommerehne, W., 177
Posner, R., 71, 86, 98
Poterba, J., 181, 191
poverty, 9, 71–2, 80, 150–1, 160–1, 171,
  193, 207, 210, 212, 228
Prein, G., x, 6, 7, 192, 193, 209
price
  discrimination, 29
  fixing, 105
prison, *see* imprisonment
Prison Reform Trust, 8, 10
prisoner's dilemma, 86
probation, 24–5, 33, 37–8, 49, 78, 151,
  169, 172, 203–5, 243–4, 247–8
property
  crime, *see* crime, property
  values, 8
prostitution, 17, 48, 50, 104
Pudney, S., 92, 96, 98
punishment, 1–3, 9, 13, 15–16, 18,
  21–2, 24–6, 28–9, 34, 71–3, 126,
  225
  capital, 37, 61, 69, 77–80, 225

Pyle, D., x, 3, 6, 74–7, 79–80, 82–3, 88,
    92, 93, 96–9, 213, 222

Quicker, J., 167, 175
quota agreements, 105

race, 15, 38, 78, 160, 161, 163, 183,
    186–7, 232
    see also ethnic
Radzinowicz, L., 60, 61, 68, 144, 148
rape, 15, 29, 31–2, 39, 46, 49, 77, 153,
    161, 228, 230, 238, 249
Rasmussen, D., x, 8, 225–6, 229, 232,
    237
rational
    choice, 14, 87, 122, 133, 210, 217–19
    maximiser, *see* maximiser, rational
rationality, bounded, 123
Raymond, J., 173
recession, 76, 94, 145
recidivism, 168–9
regional difference, 5, 7, 74, 92, 161,
    209–10, 215, 217
regression
    multiple, 91
    technique, 74, 75, 158, 159, 232
Reilly, B., 76, 81, 92, 93, 98, 212, 213,
    222
relative deprivation, 215
rent-seeking behaviour, 71
repeat victimisation, *see* victimisation,
    repeat
resource allocation, 7, 225
returns to crime, returns from crime, 3,
    82, 124, 149, 165, 170, 172, 225,
    252
return, marginal, 8, 170
Reuter, P., x, 3, 4, 10, 99, 100, 101, 102,
    104, 107–9, 111, 113, 116, 120–1,
    170, 175
reward, 2, 129, 131–6, 142, 168, 218,
    252
risk, 2, 6–8, 14, 23, 28, 45, 49, 86, 88–90,
    121–2, 124, 126–31, 133–6, 142,
    165, 172, 225
Rizzo, M., 228, 237
robbery, 16, 31–3, 46, 70, 73–5, 77, 85,
    87, 93, 143–4, 170, 195, 203,
    211–13, 217, 229–30, 249

Robinson, W., 219, 222
Rohwer, G., 195, 207
Rosenfeld, R., 211, 222
Rossi, P., 179, 191
Roth, J., 189
Rubinstein, J., 109
Russell, S., 159, 174

Sabbagh, D., 93, 98, 158, 174, 213, 221,
    251
Sacredote, B., 236, 237
Sampson, R., 168, 169, 175, 196, 208
Scheinkman, J., 236, 237
Scherer, F., 110, 120
Schilling, T., 68
Schmidt, P., 179, 180, 186, 188, 191
Schneidler, D., 190
Schumann, K., 208
Schumpeter, W., 41
self report, self report study, 163, 194,
    219
Sellin, T., 163, 182, 191
sentencing, 5, 75, 80, 149
    packages, 244
Seus, L., x, 6, 7, 192, 193, 208, 209
sexual offence, sex crime, *see* crime,
    sexual
Shapland, J., x, 8, 238, 239, 242, 246,
    250
Shavell, S., 71, 81
Shawness, L., 68
Sheley, J., 214, 221
Shipley, B., 173
Short, J., 96, 98
Sickles, R., 165, 174, 179, 188, 190
situational crime prevention, *see* crime
    prevention, situational
Sloan-Howitt, M., 131, 137
Smigel, A., 23, 28, 30, 46, 60, 68
Smith, Adam, 83, 98
social exclusion, 6–7, 9, 192–3, 200, 206
social history, 99
social loss, 2, 13, 16, 34, 36, 38, 48–9, 85
Sorsby, A., 239, 250
St Ledger, R., 148, 173, 190, 221
Steffensmeier, D., 94, 97, 213, 220
Steinert, H., 202, 203, 207
Stern, N., 73, 80, 92, 93, 97, 212, 220
Stigler, G., 47, 66, 68, 102, 118, 120

Stigler–Peltzman theory of state regulation, 118
stolen goods, handling of, 7
structural, 2–4, 139, 141, 149, 211, 214, 220
  path model, 168
supply, 2, 34, 90, 97, 108, 115, 131, 168
Sutherland, E., 60, 63, 68
Swanson, G., 180, 191
Sykes, G., 159, 173
syndicates, syndicated crime, 3, 48, 99
  see also crime, organised
system, 2, 7, 223, 225, 238, 240–1, 249

Tabbush, V., 180, 191
Tardiff, K., 211, 222
target hardening, 128–9, 236
Tarling, R., 93, 98, 213, 214, 222
Tauchen, H., x, 5, 6, 7, 163, 175–7, 182–3, 185, 191–2, 194–5, 208
Taylor, C., 166, 175
Taylor, J., 77, 79
Thaler, R., 228, 237
theft, 7, 17, 19, 29, 31–2, 46, 49, 59, 73–4, 76–7, 83, 86, 93, 127, 131, 143–5, 153–4, 159, 161, 170, 205, 209–10, 213, 215, 217–18, 228
Theil, H., 186, 191
theory of the firm, 3, 99
Thomas, D., 96, 98, 144, 148, 210, 222
Thornberry, T., 163, 168, 175, 179, 180, 191, 214, 222, 251
Timbrell, M., 213, 222
time-series, 7, 74, 76, 78–9, 96, 149, 157, 159, 209, 212–15, 251
Tonry, M., 189
tort, 40–1, 43, 70–1, 85–6
Townsend, P., 215, 222
traffic, drug, *see* drugs
trend analysis, 5, 141, 241
Trumbull, W., 165, 172, 175
Tullock, G., 71, 81

underclass, 9, 150, 194, 199
unemployment, 3–7, 9, 69–71, 75–6, 78, 82, 87–9, 93–4, 141–2, 145–51, 157–61, 163–5, 169, 171–2, 192–201, 205–7, 209–10, 212–19, 251–2

unskilled worker, 5, 6
utility, 2, 72, 123, 131–2, 172, 180
  expected, 2, 14, 22–3, 73, 87–8, 181, 218
  marginal, 124–5
utility maximiser, *see* maximiser, utility

van Dijk, J., 122, 127, 135, 137
Van Duyne, P., 4, 10
vandalism, 17, 145
victim
  compensation, 70, 86, 225
  cost, 7–8, 225–6, 229–30
  survey, 83, 153–4, 225–6
victimisation, repeat, 4, 121–3, 133–4, 136
Vigil, J., 166, 172, 175
Viscusi, W., 165, 170, 175, 179, 180, 186, 188, 191, 252
Vishen, C., 189
vocational
  qualifications, 6, 194, 196
  training, 5–6, 193–4, 201
von Mayr, A., 142, 148
Von Neumann–Morgenstern decision maker, 180
Votey, H., 181, 191

wages, 5, 59, 84, 87, 142, 159, 164–5, 169, 171, 176–7, 179, 182, 186–7, 211, 217, 227
Walmsley, J., 77, 81
Weiss, J., 163, 174
welfare, 4–5, 9, 25, 34, 46
West, D., 148, 173, 190, 221
Wheeler, D., 211, 220
white-collar crime, 16–17, 19, 46
Wiersema, B., 227, 237
Wild, R., 250
Williams, K., 211, 222
Williams, T., 166, 175
Willis, K., 74, 76, 81, 92–3, 98, 212, 222
Wilson, J., 1, 8, 10, 72, 81, 150, 170, 172, 175, 189, 191
Withers, G., 75, 76, 77, 81
Witt, R., x, 1, 76, 81, 92, 93, 98, 210, 212, 213, 217, 222

Witte, A., x, 5, 6, 7, 77, 81, 93, 98, 142,
  148, 163, 174, 175, 176, 177, 179,
  180, 182, 183, 185, 186, 188, 189,
  190, 191, 192, 194, 195, 208, 212,
  221, 252
Wolfgang, M., 163, 175, 182, 191
Wolpin, K., 74, 76, 79, 81, 92, 93, 98,
  212, 222
work ethic, 9, 202, 203, 205, 206
Wynn, S., 109

youth, 46, 77, 240
  services, 8, 165
  work, 9, 239
  see also juvenile
Yun, S., 166, 175

Zarkin, G., 75, 80, 188
zero tolerance, 193
Zimring, F., 172, 175
Zuehlke, T., 232, 237